Advances in Geographic Information Science

Series Editors:

Shivanand Balram, Canada
Suzana Dragicevic, Canada

G. Brent Hall · Michael G. Leahy (Eds.)

Open Source Approaches in Spatial Data Handling

G. Brent Hall
University of Otago
School of Surveying
Dunedin
New Zealand
brent.hall@otago.ac.nz

Michael G. Leahy
Wilfrid Laurier University
Waterloo, Ontario
Canada N21 3C5
mgleahy@alumni.uwaterloo.ca

ISBN: 978-3-540-74830-4 e-ISBN: 978-3-540-74831-1

Advances in Geographic Information Science ISSN: 1867-2434

Library of Congress Control Number: 2008932589

Cover design: deblik, Berlin

Printed on acid-free paper

9 8 7 6 5 4 3 2 1

springer.com

Preface

During the last several years there has been a significant coalescence of interest in Open Source Geospatial (OSG) or, as it is also known and referred to in this book, Free and Open Source for Geospatial (FOSS4G) software technology. This interest has served to fan embers from pre-existing FOSS4G efforts, that were focused on both standalone desktop geographic information systems (GIS), such as GRASS, libraries of geospatial utilities, such as GDAL, and Web-based mapping applications, such as MapServer. The impetus for the coalescence of disparate and independent project-based efforts was the formal incorporation on February 27th, 2006 of a non-profit organization known as the Open Source Geospatial Foundation (OSGeo). Full details concerning the foundation, including its mission statement, goals, evolving governance structure, approved projects, Board of Directors, journal, and much other useful information are available through the Foundation's website (http://www.osgeo.org).

This book is not about OSGeo, yet it is difficult to produce a text on FOSS4G approaches to spatial data handling without, in some way or another, encountering the activities and personalities of OSGeo. Of the current books published on this topic the majority are written by authors with very close connections to OSGeo. For example, Tyler Mitchell who is the Executive Director of the Foundation, is author of one of the first books on FOSS4G approaches ('Web Mapping Illustrated' (2005)). Another member of the Board of Directors of the Foundation, Markus Neteler, is co-author of the book 'Open Source GIS: a GRASS approach' (2007), which is in its third edition.

Hence, not coincidentally, the current text has contributions from a number of authors with close connections to the Foundation. The importance of OSGeo in all aspects of FOSS4G development is unquestionable and unparalleled in the development of Open Source software within the spatial data domain. As OSGeo has established itself firmly at the centre of what is described by David McIhagga in Chap. 3 of this book as the Open Source Web mapping ecosystem, FOSS4G technologies and concepts have permeated into many diverse areas of application development. One area where recent interest in the spirit and 'openness' of FOSS4G software is apparent is the tertiary education sector. This interest is likely to increase in the future as the tools, which are already of quite an incredible level of maturity, become better known and ever more widely used, and as curricula evolve to explore opportunities other than those that are tightly coupled with the dominant proprietary GIS software vendors.

The idea for this book evolved from working with FOSS4G tools on projects within an academic environment. Other than the two texts referred to above and one or two others, prior to the appearance of OSGeo the dominant reference source for FOSS4G projects was and substantially still is the Internet. This is perhaps the way it should be as, with projects that are in a constant state of evolution, the use of the printed word is inevitably associated with a limited shelf-life. This is especially true for texts that are 'cook book' oriented, containing instructions on how to do this or that with a specific software package. These sorts of texts are essentially only relevant to the versions of software that they relate to, yet there is a huge international market for them. Hence, the approach sought and largely implemented with this book was not to provide readers with information on how to use a specific FOSS4G tool or project, but rather to focus on several projects more from a conceptual rather than a 'how to' point of view. The purpose is to introduce readers new to FOSS4G software to the nature, purpose, evolution and characteristics of a number of projects, while also discussing important issues such as the role of standards in OS software development, the business models that can allow 'free' software development to sustain the developers, and the general need for a spirit of co-operation and partnership building that is often absent from the closed software marketplace.

The book is generally divided into three sections. The first three chapters focus on the topics noted immediately above. New business models have had to be created or have evolved to promote and sustain core FOSS4G projects, and companies, such as DM Solutions Ltd., Refractions Research and others that have grown on the back of FOSS4G inspired ideas. Hence, market niches have been identified that allow these commercial entities to provide FOSS4G services and FOSS4G solutions that remain freely accessible to any and all who are interested, while remaining commercially viable. Equally, issues such as the use of standards, improving documentation, making tools more accessible to end-users who are not programmers, improving FOSS4G interoperability, co-ordinating collaboration within the developer community, and controlling code release frequency through use of best practice management standards are all now substantially more important than beforehand, with the groundswell of support for and involvement in project development. Respectively, these three initial chapters are written by Tom Kralidis on standards, Arnulf Christl on new business models, and David McIhagga on what he aptly describes as the Open Source Geospatial Web 'ecosystem'.

The second section of the text comprises the majority of the chapters. In this section focus is given to a variety of key FOSS4G projects. For the most part the authors of these chapters are individuals who, for the most part, have been either the catalyst of the project or have played a prominent role in its development. Each chapter is generally built around a discussion of the objectives of the project, the architecture of the tool(s), how the project evolved to satisfy its initial objectives, or alternatively how the objectives morphed as the project unfolded into its current state. There is some technical discussion in these chapters, however the intention of the text was not to produce a manuscript shrouded in technical language. While this is inevitable to some extent with technical subject matter, the intention, as noted above, was to make the book accessible to those new to FOSS4G, while also providing information of interest to established members of the FOSS4G community.

The chapters are illustrated to varying extents, some richly and some not at all, with design diagrams and screen captures. Clearly, it was not possible to cover all of the core or key FOSS4G projects that have evolved, but the chapter selection does a fair job at spanning the field. In fact, chapters discuss five of OSGeo's 13 established Web mapping, desktop, or geospatial library projects. Respectively, Chap. 4 is authored by Steve Lime and focuses in MapServer, perhaps the most successful Web-based mapping tool yet developed. Chapter 5, by Frank Warmerdam, discusses the Geospatial Data Abstraction Library (GDAL), which also is possibly the most successful such library to have yet been produced. The 6th Chapter is by Rongguo Chen and Jiong Xie and it deals with Open Source databases and their spatial extensions, most notably PostGIS, produced by Refractions Research. Chapter 7 is by Robert Bray and it is important as it deals with MapGuide Web mapping software, by Autodesk Inc., which was converted to OS in November 2005. The software was contributed to OSGeo in March 2006 as a foundation project. Chapter 8 is by Ian Turton and focuses on GeoTools, which is Java-based and is used as the base for several other well known FOSS4G projects. Chapter 9 is by Markus Neteler and his colleagues from the GRASS development team, and it discusses probably the oldest and most firmly established FOSS4G desktop application available.

The third section of the book comprises the same number of chapters as the first. These chapters discuss applications of some of the tools described in section two, with the addition of the impressive array of FOSS4G libraries and applications packages developed by Gilberto Camara and his colleagues from various institutions in Brazil. Specifically, Chap. 10 discusses one of two University-based FOSS4G projects reviewed in the book, namely GeoVISTA *Studio* developed by Mark Gahegan and his colleagues at Penn State University. Chapter 11, by the Editors, discusses a second University-based tool which utilizes several of the other FOSS4G projects discussed in the text.

Compiling these chapters was an interesting exercise. It became very clear very quickly, and remained clear throughout the project, that the one element that is in very short supply in FOSS4G software development in general is free time. Hence, for several of the contributors finding time to complete their chapter proved to be difficult. In addition, it seems also that outstanding programmers enjoy doing what they do best, but do not necessarily enjoy writing about it! Given this, we would like to thank the contributors for their forbearance in my persistent 'nagging', which was required in order to get the book finished. In the final analysis, the chapters together weave a very useful tapestry of activities within this general field.

It would be remiss of us not to complete this preface without noting thanks to a number of individuals who have helped along the way. First, we would like to thank the series editors, Drs. Shivanand Balram and Suzana Dragicevic from Simon Fraser University, British Columbia, Canada, who liked the initial idea of doing a book on this theme. Chris Bendall, the editor from Springer was also very helpful in moving the idea into a reality. We would like to thank our colleagues at the University of Waterloo, Ontario, Canada, where the bulk of the editorial work for this book was done. In particular, we wish to thank Dr. Rob Feick, as well as the following graduate students who worked on the MapChat project, namely David Findlay, Taylor Nicholls, John Taranu and Brad Noble. Last but certainly not least we thank our wifes Masha and Ally, whom we dedicate this effort to.

Dunedin, New Zealand

G. Brent Hall
Michael G. Leahy

References

Mitchell T (2005) 'Web mapping illustrated', O'Reilly Media Inc., Sebastopol, Ca.
Neteler M, Helena M (2008) 'Open source GIS: A GRASS GIS approach', 3rd edn, Springer, New York

Contents

List of Contributors

D.E. Beaudette
Department of Land, Air and Water Resources, University of California, Davis, CA 95616, USA, e-mail: debeaudette@ucdavis.edu

Robert Bray
Autodesk, Inc., 2100, 645 - 7th Ave SW, Calgary, Alberta T2P 4G8 Canada e-mail: robert.bray@autodesk.com

Gilberto Câmara
National Institute for Space Research (INPE), Av dos Astronautas 1758, 12227-010, São José dos Campos, Brazil, e-mail: gilberto@dpi.inpe.br

Marcelo Tílio de Carvalho
Catholic University of Rio de Janeiro (PUC-RIO), Rua Marquês de São Vicente, 22522. 453-900 Rio de Janeiro/RJ, Brazil, e-mail: tilio@tecgraf.puc-rio.br

Marco Antonio Casanova
Catholic University of Rio de Janeiro (PUC-RIO), Rua Marquês de São Vicente, 22522. 453-900 Rio de Janeiro/RJ, Brazil, e-mail: casanova@tecgraf.puc-rio.br

P. Cavallini
Faunalia, Piazza Garibaldi 5, 56025 Pontedera (PI), Italy,
e-mail: cavallini@faunalia.it

J. Cepicky
Help Service - Remote Sensing s.r.o., Cernoleska 1600, 25601 - Benesov, Czech Republic, e-mail: jachym.cepicky@gmail.com

Rongguo Chen
State Key Laboratory of Resources and Environmental Information System, Chinese Academy of Sciences, Beijing, China, e-mail: chenrg@lreis.ac.cn

Arnulf Christl
WhereGroup GmbH & Co. KG, Siemensstr. 8, 53121 Bonn, Germany; Open Source Geospatial Foundation, Beaverton, OR, USA,
e-mail: arnulf.christl@wheregroup.com, arnulf@osgeo.org

Urška Demšar
National Centre for Geocomputation, National University of Ireland, Ireland,
e-mail: urska.demsar@nuim.ie

Karine Reis Ferreira
National Institute for Space Research (INPE), Av dos Astronautas 1758, 12227-010, São José dos Campos, Brazil, e-mail: karine@dpi.inpe.br

Ubirajara Moura de Freitas
Space Research and Applications Foundation (FUNCATE), Av. Dr. João Guilhermino, 429 – 18th floor 12210-131 São José dos Campos, SP, Brazil,
e-mail: bira@geo.funcate.org.br

Mark Gahegan
Department of Geography, University of Auckland, Private Bag, Auckland, New Zealand, e-mail: mark@geog.psu.edu

G. Brent Hall
School of Surveying, University of Otago, Dunedin, New Zealand,
e-mail: brent.hall@otago.ac.nz

Frank Hardisty
GeoVISTA Center, Department of Geography, Penn State University, Pennsylvania, USA, e-mail: hardisty@psu.edu

Athanasios Tom Kralidis
Open Source Geospatial Foundation, Beaverton, OR, USA,
e-mail: tom.kralidis@gmail.com

L. Lami
Faunalia, Piazza Garibaldi 5, 56025 Pontedera (PI), Italy,
e-mail: lami@faunalia.it

Michael G. Leahy
Department of Geography and Environmental Studies, Wilfrid Laurier University, Waterloo, Ontario, Canada, N2L 3C5, e-mail: mgleahy@alumni.uwaterloo.ca

Stephen Lime
Minnesota Department of Natural Resources, St. Paul, MN, USA,
e-mail: steve.lime@dnr.state.mn.us

David McIhagga
President & CEO, DM Solutions Group, Ottawa, ON, Canada,
e-mail: dmcilhagga@dmsolutions.ca

Antônio Miguel Vieira Monteiro
National Institute for Space Research (INPE), Av dos Astronautas 1758, 12227-010, São José dos Campos, Brazil, e-mail: miguel@dpi.inpe.br

M. Neteler
Fondazione Mach - Centre for Alpine Ecology, 38100 Trento (TN), Italy, e-mail: neteler@osgeo.org

Gilberto Ribeiro de Queiroz
National Institute for Space Research (INPE), Av dos Astronautas 1758, 12227-010, São José dos Campos, Brazil, e-mail: gribeiro@dpi.inpe.br

Ricardo Cartaxo Modesto de Souza
National Institute for Space Research (INPE), Av dos Astronautas 1758, 12227-010, São José dos Campos, Brazil, e-mail: cartaxo@dpi.inpe.br

Masa Takatsuka
ViSLAB, Information Technologies, University of Sydney, Australia, e-mail: masa@vislab.net

Ian Turton
GeoVISTA Center, Pennsylvania State University, University Park, PA, 16802, USA, e-mail: ijt1@psu.edu

Lúbia Vinhas
National Institute for Space Research (INPE), Av dos Astronautas 1758, 12227-010, São José dos Campos, Brazil, e-mail: lubia@dpi.inpe.br

Frank Warmerdam
Independent Software Developer, Open Source Geospatial Foundation, Beaverton, OR, USA, e-mail: warmerdam@pobox.com

Jiong Xie
State Key Laboratory of Resources and Environmental Information System, Chinese Academy of Sciences, Beijing, China, e-mail: xiej@lreis.ac.cn

Chapter 1
Geospatial Open Source and Open Standards Convergences

Athanasios Tom Kralidis

Abstract Geospatial information has become a ubiquitous resource. The proliferation of the Internet and information technology has resulted in an enormous volume of information exchange and a growing global geospatial data infrastructure presence. Interoperability is increasingly becoming a focus point for organizations that distribute and share data. Standards are an essential aspect of achieving interoperability. This chapter illustrates the benefits of using open standards for geospatial information processing. It also discusses various free and open source for geospatial software (FOSS4G) packages that support open standards. Finally, the chapter illustrates how open source software and open standards can be easily integrated in a number of scenarios.

1.1 Introduction

Laying the groundwork to establish a framework for the interoperability of spatial data has been an ongoing activity for at least three decades. The 1970s saw the emergence of a growing requirement for national mapping and surveying agencies to create policies, agreements and processes for normalizing access to and applications of spatial data. In Canada, for example, the origins of a spatial data infrastructure emerged in the 1980s as an effective means of facilitating data access (Groot and McLaughlin 2000). Following from this there has been an ongoing effort worldwide to produce standards-based specifications for the discovery, evaluation, access, visualization and exploitation of spatial data resources (Global Spatial Data Infrastructure Association 2001).

This chapter discusses the concept of interoperability, the roles and activities of open standards bodies and organizations, and provides examples of free and open source software for geospatial (FOSS4G) projects which exemplify standards-based approaches to spatial information exchange and processing.

Athanasios Tom Kralidis
Open Source Geospatial Foundation, Beaverton, OR, USA, e-mail: tom.kralidis@gmail.com

G.B. Hall, M.G. Leahy (eds.), *Open Source Approaches in Spatial Data Handling.*
Advances in Geographic Information Science 2,

1.1.1 Geographic Information

Over the last three decades, governments and industry have invested billions of dollars in the development of geographic information systems (GIS) to serve various information communities including forestry, marine studies, disaster management, natural resources, health, and numerous others (Groot and McLaughlin 2000). The information collected by organizations from within these communities has the potential for multiple uses and sharing between users, activities, systems and applications. Despite significant decreases in the cost of computer hardware and software over time, spatial data are ever more voluminous and an expensive resource to develop and maintain. One means that has become popular to organize spatial data resources is the concept of a spatial data infrastructure.

1.2 Spatial Data Infrastructure

This section discusses the concept of a spatial data infrastructure (SDI) by focusing on digital networks and the Internet as the foundation global data infrastructure and discusses how SDIs leverage the presence of the Internet to establish data sharing and data exchange mechanisms. This discussion also illustrates how the concept of interoperability drives the functionality of an infrastructure and is a core requirement for information exchange of any kind.

1.2.1 The Internet and the Digital Age

A data infrastructure can be defined as a transparent, robust computer environment, which enables access to information using common, well-known and accepted specifications, standards and protocols (Global Spatial Data Infrastructure Association 2001). To use a simple analogy, a telephone network can be thought of as an interoperable infrastructure, in that it provides users with connectivity and services to communicate with each other, while the details behind the communications, including the physical telecommunications infrastructure such as networks, wiring, switches, and exchanges are transparent and relatively unimportant from an end user perspective. Such an infrastructure can be seen as an underlying building block to enable communications by products such as specialized applications, as well as the development of sub-networks to be built and deployed for specific purposes. Although a critical aspect of networking and communications, this form of interoperability is also mundane in its ubiquity. However, the very existence of the infrastructure required to facilitate communications makes enabling objects, technologies, and analysis possible (Harvey 2000).

A data infrastructure is also the result of many nodes around which data and services are decentralized. This process of dispersion of data and service points on a densifying network is similar to the process that organizations go through when they are restructuring from single site to multi-site enterprises. For example, in the 1980s and 1990s, globalised forms of enterprise organization began to emerge where companies sought to leverage lower costs. To a large extent, this process was facilitated by the growth of network-based forms of enterprise organization as well as information sharing. Though location is an important consideration in this process, the decentralization of economic activity has shown that location is not as important in the context of doing business as it once was. This principle also applies to infrastructures that gather, process, and disseminate geographic information. Organizations can collect and maintain their own data holdings, and publish them through clearinghouses for use by end users and/or an ever widening variety of data services. In this process data are kept closest to the source of production or collection to facilitate their update and completeness, but equally they may become widely dispersed geographically in their use and potentially also in their enhancement.

A SDI extends the data infrastructure concept by focusing on the transport and transmission of spatial information and in providing the relevant technologies, policies and agreements that assist in the availability of and access to spatial data resources. A SDI provides the architectural underpinnings for the discovery, evaluation and application of spatial information (Global Spatial Data Infrastructure Association 2001). In this context, a concerted effort is currently being made among government agencies, both within and between countries, to enable the discovery, visualization and access of spatial information at all levels by leveraging the Internet as the distributed infrastructural backbone for interaction with spatial data (Canadian Geospatial Data Infrastructure Architecture Working Group 2001; Global Spatial Data Infrastructure Association 2001). Examples of organizations supporting such activities include the United States Government's Federal Geographic Data Committee (FGDC – Federal Geographic Data Committee 2004), the National Aeronautics and Space Administration (NASA – National Aeronautics and Space Administration 2004) and Canada's GeoConnections program (GeoConnections 2007). Other examples include the Australian and New Zealand Land Information Council (ANZLIC – Australian and New Zealand Land Information Council 2007), and the Infrastructure for Spatial Information in Europe (INSPIRE – Inspire 2007).

Using the infrastructure approach to geographic information management, organizations can interact with spatial data over the Internet in a transparent fashion. A SDI reduces the requirements for multiple standards by establishing a unified approach to data syntax, semantics and schemas as well as content management. Moreover, a SDI encompasses networked spatial databases and allows efficient management of complex organizational, technical, human and economic components, all of which interact with one another. Without this centralized and unified approach to management, the cost and feasibility of multiple copies of spatial information quickly become unmanageable, especially if current data are an important requirement for an enterprise. Locally stored copies of data result in large, ongoing data management problems for organizations. In addition they are costly to sustain, and are prone to concurrency issues with multiple versions of the same data set existing within and between the same and different organizations.

Computer networks are vital to a SDI (Groot and McLaughlin 2000). Most distributed network systems are made up of client-server architectures, where one or more central servers provides services and information to client via an information exchange mechanism, which is vital to the infrastructure the network supports. The Internet, which is built upon this model of computing, originated from the United States Department of Defense Advanced Research Projects Agency Network (ARPANET) in the late 1960s (Begg and Connolly 1998). Its current popularity has reinforced and generated numerous diverse information highways in many information communities (Hartman 1997). Computers now leverage network technology to share disk drives, memory, and/or data. The Internet and the TCP/IP reference model, provides a means for transporting information "packets", providing a framework (network addressing, fragmentation, timeouts, and so on) for peer-to-peer communication through TCP, and enabling an application layer for user-level protocols, such as File Transfer Protocol (FTP) and Hypertext Transfer Protocol (HTTP) (Groot and McLaughlin 2000).

The Internet enables information holdings and services to be distributed in terms of their location. Based on open communication standards and protocols, this has enabled organizations to publish information over a distributed network infrastructure, as well as provide a medium for discovering and evaluating educational resources, commercial initiatives and government information among other things. The common standards allow computers to connect to the backbone network and to each other despite differences in hardware, software and other factors that have historically impeded communications (Hartman 1997). This results in a foundation layer of interoperability in network communications.

1.2.2 Interoperability

Interoperability can be defined as the ability of a system or components of a system to provide information sharing and inter-application process control, through a mutual understanding of request and response mechanisms (Groot and McLaughlin 2000). Interoperability is the ability of a system (or a component of a system) to access a variety of heterogeneous resources by means of a single, unchanging operational interface (Canadian Geospatial Data Infrastructure Architecture Working Group 2001). That is, two resources (such as a client and a server) are interoperable if there is a mutually agreed upon and standardized messaging vocabulary which they can understand. While communications may relay different requests and responses, the two resources understand the frameworks in which they are delivered.

The concept and practice of interoperability dovetails with the open systems model, which is an approach to software engineering and system design that enables and encourages the sharing of resources (Gardels 1999; Guerrero 2004). These resources are regarded as objects, meaning that every resource can be seen as a component among other components which coexist under a common framework, thus promoting an operational model as opposed to data standards (Gardels 1999). Such common frameworks benefit from publically available and agreed upon methods and practices for information processing within and between various communities and networks.

1.3 Open Standards

This section provides an overview of open standards to support information processing in the geospatial domain, from what they are, to identifying key formal standards bodies as well as ad hoc and de facto standards groups. Current adoption and organizational benefits of using standards are also discussed to illustrate the significant influence open standards have had and will continue to have in the FOSS4G arena. Though there are numerous standards and standards bodies, those with particular relevance to the FOSS4G community are the focus of interest.

1.3.1 Overview

The ubiquity of geospatial information results in massive information repositories maintained by mapping and surveying organizations that publish content to SDIs. As suggested earlier, the Internet has had an enormous impact in enabling the discovery, access and visualization of spatial data, for both information providers and consumers alike. The Internet has provided the ability to integrate data holdings, and provides a transparent layer to the end user to interact with spatial data resources (Begg and Connolly 1998). With the advances in computer technology and standards, SDI activities increasingly provide an opportunity for the cost-effective collection, sharing and distribution of information with a geographic component within and between user communities (Groot and McLaughlin 2000). With the volume of spatial data being produced and published to the Internet ever increasing, issues emerge with regard to usability and suitability. For example, the following questions, among others, must be answered adequately:

- Are the spatial data posted on the Internet in a format or structured in a way in which those wishing to utilize the data can comprehend and interpret them relative to application and analytic needs?
- Do the data originate from an authoritative and reliable source or provider?
- Are the data representative of the most current updates and maintenance by the authoritative provider?
- Is the consumer looking for an entire data product, or for a specific parcel, region, or band combination of imagery? That is, the consumer may be seeking a subset of a much larger database, but cannot afford to, or may not wish to, acquire the entire data collection.
- Do the data have any security and/or policy issues with regard to their usage?

These questions represent just a few issues with regard to spatial data interoperability within a SDI. The causes of such issues can be due to differing organizational policies and practices, as well as contradictory approaches to information management, technology and data sharing within the spatial data community. In each case it is likely that the lack or ineffective use, of specifications and standards is at the heart of difficulties that are encountered (Groot and McLaughlin 2000).

Harmonizing approaches and standards for spatial data acquisition and exchange lessens the requirement for maintaining multiple versions of the same data, publishing the data, and exchanging data within and between provider and user groups, all of which may become very expensive in terms of resources and operating budgets. The 1970s saw the emergence of a growing requirement for national mapping and surveying agencies to create policies, agreements and processes for normalizing the access to and application of spatial data resources (Groot and McLaughlin 2000). These requirements were initially narrow in scope and have increasingly come to focus on the use of data standards.

A standard can be best defined as a document or collection of documents, usually but not always published, that establish a common language, terminology, accepted practices and levels of performance, as well as technical requirements and specifications, that are used consistently for the development and use of products, services and systems (Yeung and Hall 2007). Standards are multidimensional. That is, they can be defined for data content, values and at various levels of conformance, such as technical specifications, conventions, and guidelines.

Standards initially provide three primary benefits for spatial data and their users, namely portability, which includes use and reuse of information and applications; interoperability, which includes multiple system information exchange; and maintainability, which includes long term updating and effective use of a resource (Groot and McLaughlin 2000). Standards can save time and effort by removing the need for reinventing approaches to discovery, evaluation, access and visualization of spatial data. Standards organizations for SDI are evident at multiple levels, such as government organizations (for example the FGDC), independent bodies, such as the Canadian General Standards Board (CGSB), American National Standards Institute (ANSI), the International Organization for Standardization (ISO), and industry associations, such as the Open Geospatial Consortium (OGC) (Groot and McLaughlin 2000). A SDI supports low-level standards, such as computer hardware, networks and operating systems, as well as high-level standards such as user interfaces, data formats, and presentation views of data (United States National Research Council 1999).

Standards promote interoperability within an infrastructure, and provide significant benefits for information exchange. Standards are designed for broad, long-term use. However, they are not immutable, and may be modified by consensus among users by standards-issuing bodies. This process may occasionally pose difficulties due to the lengthy design and definition process used to create a standard, which initially takes a potentially long process of submission and review iterations before the standard is accepted by the body it relates to. Once the standard is approved, it is up to the relevant organizations or communities of users to utilize its content.

The above mentioned enabling approaches and technologies provide open-ended possibilities for geospatial information. However, they also raise issues of data copyright and intellectual property (IP). Open standards are independent of IP and organizational policies regarding spatial data in that they can be applied to any IP/policy situation for spatial data. The development of a useful legal framework for both private and public activity is vital to the dissemination of spatial data, no matter what standard is used. As noted earlier, geographic information is not cheap to produce and maintain (Aslesen 1998). The capabilities of digital infrastructures and information communities create further concerns over geospatial information and its potential misuse as control over copying data is difficult to implement. In fact, GIS and related technologies can be dangerous in their ability to merge spatial data by identifying details and information that are otherwise not transparent independently.

Hence, it is not surprising that there is a community of interest on matters relating to SDI in general and the role of enabling standards in particular. How does this community communicate in terms of discovering, evaluating, accessing, and visualizing spatial information? How is interoperability prescribed in a SDI, and how does it satisfy the requirements of the community? These questions and other related issues are discussed further below.

1.3.2 Relation to Open Source

In this discussion it is important to distinguish between open standards and open source in order to be aware of differences in the meanings of these terms, as they are sometimes used interchangeably. An open standard (i.e. a standard that is publicly available to use) can be implemented by open source (i.e. the principles and methodologies to promote open access to design and production) software, as well as commercial or proprietary solutions, in much the same manner. However, open standards are implementation agnostic and are not exclusive to open source software.

1.3.3 World Wide Web Consortium (W3C)

Established in 1994, the World Wide Web Consortium (W3C) develops interoperable technologies (specifications, guidelines, software, and tools) to facilitate minimal levels of conformity in Web standards. The Consortium describes itself as a forum for information, commerce, communication, and collective understanding (World Wide Web Consortium 2007). W3C standards are free to obtain and implement. Core specifications include Hypertext Transfer Protocol (HTTP), the Uniform Resource Locator (URL), and Hypertext Markup Language (HTML), which have become building blocks for other W3C specifications such as the Document Object Model (DOM), Extensible Markup Language (XML), Extensible Stylesheet

Language Transformations (XSLT), Scaleable Vector Graphics (SVG), and Cascading Stylesheets (CSS). More information on these specifications is available at http://www.w3.org/.

1.3.4 The International Organization for Standardization (ISO)

The ISO is an international standard-setting body composed of representatives from various national standards bodies. ISO was founded in 1947 to produce world-wide industrial and commercial standards (International Organization for Standardization 2007a). Many ISO standards become nationally endorsed, and are heavily used and implemented for a variety of areas, such as the ISO 9000 series of quality management standards (International Organization for Standardization 2007a).

Within the ISO there are numerous technical committees (TCs). TC211 is the committee responsible for standards pertaining to digital geographic information (International Organization for Standardization 2007b). TC211 produces a number of abstract specifications and reference models. These specifications are typically used as building blocks for other standards to leverage, such as those from the OGC. At the time of writing, current implementation specifications of interest include 19115, which covers digital geospatial metadata (International Organization for Standardization 2007c), and 19138, which is the implementation standard of 19115 (International Organization for Standardization 2007d).

1.3.5 The Open Geospatial Consortium (OGC)

The OGC was founded in 1994 as the OpenGIS Consortium. It is a non-profit, international, voluntary consensus standards organization specializing in geospatial data and Web services. The OGC consists of over 250 organizations from government, academia, industry and other groups. The Consortium was founded on the concept of providing open specifications at no cost to the public to acquire and/or implement, thus providing standards-based interfaces for geospatial discovery, access, visualization and processing. The OGC leverages existing efforts from other standards organizations such as the W3C, ISO, and the Organization for the Advancement of Structured Information Standards (OASIS), and builds upon them in reference to the spatial data domain.

The OGC Abstract Specification provides the reference model for implementation of OGC specifications. Areas covered by the Abstract Specification include:

- Feature Geometry
- Spatial Referencing by Coordinates
- Locational Geometry Structures
- Stored Functions and Interpolation
- Features
- The Coverage Type

- Earth Imagery
- Relationships Between Features
- Feature Collections
- Metadata
- The OpenGIS Service Architecture
- Catalog Services
- Semantics and Information Communities
- Image Exploitation Services
- Image Coordinate Transformation Services
- Location Based Mobile Services
- Geospatial Digital Rights Management Reference Model (GeoDRM RM)

The OGC Abstract Specification represents a carefully engineered process and framework in support of the discovery, access and visualization of spatial data. Specifications, discussion papers and recommendation papers are developed from the vision of the Abstract Specification. For example, all OGC Web Services (OWS) provide models for metadata documentation. The Geography Markup Language (GML) specification models spatial features and topological relationships between them. The widely used Web Feature Service (WFS) describes a service-based supply of vector information as feature collections. All OWS follow the Service Architecture Interoperability approach.

The OGC has a strong and progressive specification development process that requires consensus between specific working group members. OGC specifications are typically developed, tested and revised within the OGC testing environments (or "testbeds"), pilot projects, and working group activities. The major benefit of this approach is the iterative process of standards development in concert with specification and software development. Typical specification development takes place by defining, adopting, and publishing the specification document for vendors and others to implement. It is common practice for specifications not to take into account various aspects which may affect ease of software development, functionality and/or usability.

The result of this is often a revision process which can become resource intensive and time inefficient. Vendors may subsequently add "vendor specific" functionality to software, which is where variations begin to surface across vendor implementations of the same specification. In the OGC environment, because the specification is developed with software implementers, this risk is significantly reduced, allowing for specifications to be tested, analyzed, and updated before they reach public adoption. The result is a stronger, more robust version of a given specification and multi-vendor interoperability of software products. This approach to specification development also introduces the unique concept of competing businesses in the spatial information industry working together in a co-operative manner. A full listing of publicly adopted OGC specifications can be found at http://www.opengeospatial.org/standards.

Since the Web Map Service (WMS) was published in 1999 as the first major OGC specification, the OGC has gained a great deal of momentum and credibility in terms of organizational recognition, resulting in many early adopters of geospatial Web services and interoperability. In fact, a survey in 2004 indicated 166 public OGC WMS instances found via the Google search engine. The survey, though not authoritative or scientific, uses Web development to collect and provide reports on OGC usage over the Internet. While the specific searching and interpretation algorithms of the survey remain subject to further interpretation, it is evident that the number of servers indicate a level of maturity and popularity with the OGC and Web service approaches. OGC instances were found in Canada, the United States, Germany, Netherlands, Australia, Italy, Denmark, Czech Republic and Mexico (Ramsey 2004).

The OGC specifications are also making their presence felt in major GIS vendor software packages. This can also be attributed to industry recognition and in response to organizational requirements based on the underlying benefits of interoperability and the Web services approach. In 2007, 381 vendor products either implement or directly conform to OGC specifications (Open Geospatial Consortium 2007a).

Current OGC activities of interest include Sensor Web Enablement (SWE), which involves defining specifications for sensor-based instruments and platforms, Digital Rights Management, and approaches for more complex geoprocessing (Web Processing Service) and linking (Geospatial Linking) (Open Geospatial Consortium 2007b,c). The OGC liaises with ISO through a cooperative agreement where both organizations can leverage and align with one another's developments while satisfying organizational requirements (International Organization for Standardization 2007e), and recognizing and leveraging other standards groups such as the W3C.

In addition to the above specifications, Keyhole Markup Language (KML) Version 2.2 was adopted as an OGC standard in April 2008 (for more information see http://www.opengeospatial.org/standards/kml). This format allows for visualization in applications such as Google Earth, Google Maps and Google Maps for Mobile (Google 2007). KML is an XML-based grammar which has gained a great deal of popularity and is now used in many applications that run over the Internet supporting data styling and referencing in a single document.

1.3.6 De Facto and Ad Hoc Standards

In addition to formal standards bodies and specification programs, there exist numerous de facto standards, which are illustrative of being developed in a relatively informal setting (mailing lists, wikis, forums, etc.), and mature (somewhat organically) to become so popular that they are followed as if they were formal standards.

Some mainstream Web examples include JavaScript, which was originally developed by Netscape, and subsequently became ECMAScript in DOM 1 and 2 HTML (Flanagan 2006). Perhaps the most popular example of this in the spatial data domain is the shapefile (Environmental Systems Research Institute 1998). Since the inception of this format in ArcView software, most GIS packages have created support to read and write this relatively simple form of spatial data encoding. Other de facto standards include the following developments:

- GeoRSS – This extends Really Simple Syndication (RSS) feeds with location information. GeoRSS has become very popular and is implemented by major applications such as Google Maps, Yahoo Maps, and a variety of other implementations. GeoRSS has since been released as an OGC White Paper (Open Geospatial Consortium 2006)
- GeoJSON – Extends JavaScript Object Notation (JSON) to encode objects with location information expressing a variety of geographic data structures (JSON 2007). JSON is a widely used approach for working with data in rich, Web 2.0 style applications, which can instantiate JavaScript objects directly without transforming them from an intermediary format
- Tiled Maps – This approach replaces arbitrary resolution Web mapping approaches with "tiled" maps, which can be managed in an underlying cache mechanism by software (Open Source Geospatial Foundation 2006a,b,c). This approach has allowed the development of rich Web mapping applications that are very responsive relative to the earlier means of rendering these maps in Web-based mapping. Efforts have resulted in a de facto Tiled Map Service Specification (TMSS), and WMS Tiling Client Recommendation (TCR).

It is evident that there are many standards bodies and options that deal directly or indirectly with various aspects of spatial data. It is important to understand the roles of these bodies (e.g., W3C as pure Web-oriented, OGC as spatial data-oriented) to assess what standard is best suited for a given requirement. It is also encouraging that standards bodies have been and are increasingly working with one another so as not to duplicate effort and maximize leverage of accepted approaches.

1.3.7 Current Adoption and Organizational Benefits

It is evident that open standards are making an impact in the geospatial community. As previously mentioned, OGC specifications are supported across numerous software packages (both desktop and Web-based). De facto geospatial standards such as the shapefile and GeoRSS have also gained much attention and are being adopted by the mainstream IT community.

Extended collaboration and partnerships using open standards provide organizations with the opportunity to create open interfaces and communication mechanisms for distributed computing. In the absence of open standards, application client software packages are "bound" to the interfaces and operations as prescribed by the organization or service provider. The result of this is that whenever some aspect of business logic or process is modified at the service level, clients must align with those changes to ensure the same level of service and information is maintained.

Using open standards also lowers the barrier to integration. That is, well known standards can foster the development and use of common tools and technologies, which can act as building blocks for developers. For example, a Java developer with spatial data requirements in his/her existing application can leverage a package like

GeoTools (GeoTools 2007) instead of implementing something from scratch, allowing more time for including other resources in their business domain requirements.

1.4 Open Source Standards-Based Examples

There already exist many open source tools and technologies which implement and/or conform to open standards. This section outlines several of the FOSS4G packages from various parts of the value chain (e.g., servers, clients, databases), and discusses how the existence of open standards have benefited the development and maintenance of these software packages.

1.4.1 MapServer

MapServer is illustrative of a FOSS4G software package which heavily implements open standards. Originally developed as a C language Web mapping engine with a common gateway interface (CGI), MapServer also provides a scripting environment with bindings to popular scripting languages (perl, php, python, ruby, java, c#, tcl, etc.) for easier integration into scripting environments.

MapServer initially supported OGC WMS version 1.0.0 in 2001 (Regents of the University of Minnesota 2001). At the time of writing, OGC support includes WMS 1.1.0 and 1.1.1, WFS, GML, Web Coverage Service (WCS), Sensor Observation Service (SOS), Web Map Context (WMC), Styled Layer Descriptor (SLD), Filter Encoding Specification (FES), and OWS Common.

As standards emerge and evolve, MapServer continues to respond to requirements for open standards support and implementation, which exemplifies convergence between open standards and open source. As a result, MapServer software can communicate with any other (open source or commercial) software package via standards based interfaces and encodings, as either a publisher or consumer in a client-server scenario. More in depth information on MapServer can be found in Chap. 4 of this book and at the MapServer Web site (Regents of the University of Minnesota 2007).

1.4.2 Community Mapbuilder

Community Mapbuilder is a powerful, Web 2.0 style, standards-compliant geographic mapping client which runs in a Web browser. Mapbuilder is a pure browser-based solution. That is, all code operates on a Web browser client (such as Firefox, Internet Explorer) as HTML and JavaScript. No special plug-ins or browser extensions are required, hence Mapbuilder can be classified as a "thin client" or AJAX (Adaptive Path 2005) solution.

Mapbuilder is a strong proponent of open standards, using XSLT as the core XML processing functionality, and using WMC (hence WMS) at the core of its mapping display capabilities. It also supports WFS, WFS-T, GeoRSS, SLD, GML and more. Heavy use and implementation of standards are a main reason of Mapbuilder's progressive development, which gained momentum in 2004 (Community Mapbuilder 2007).

Despite this promise, and perhaps indicating the general nature of OS development, the Mapbuilder Steering Committee announced at the end of July 2008 that after the release of Version 1.5 of the software it would be retired and there would be no planned enhancements to it. The rationale for this relates to the growth of what is described in Chap. 3 as the Web mapping ecosystem. Specifically, the Project Steering Committee refers to the growth of interest in the Openlayers project (http://openlayers.org) and the resulting diversion of users and developers from Community Mapbuilder to Openlayers. Despite this, and the formal end of life of the project, the beauty of OS means that the code is still alive and that independent developers can continue enhancing it for as long as they like.

1.4.3 PostGIS

PostGIS provides support for geographic objects to extend the PostgreSQL object-relational database software package (see Chap. 6 of this book). PostGIS is a C-based spatial engine which implements the OGC SFS specification (Refractions Research 2005). The benefits of SFS allow for standards-based support of spatial processing functions, such as whether two geometries overlap, are within one another, and so on. Also valuable is the support for input and output of the OGC's Well Known Text (WKT) format, as shown later in this chapter.

As PostGIS can serve as a backend to other FOSS4G packages such as MapServer, GeoServer, uDig, and GRASS among others, the OGC support is further exposed to calling codebases. For example, a C developer connecting to the PostgreSQL C API can make direct SQL calls using PostGIS asgml() functionality to retrieve spatial data encoded as GML.

1.4.4 Others

Numerous other geospatial software packages support open standards. For example, GeoServer is a Java-based geographic server which supports WMS and WFS, as well as WFS-T. uDig is a desktop Internet-enabled GIS which supports WMS, WFS, and other related standards. Geonetwork Opensource supports the Catalogue Services specification as well as ISO 19139 for metadata. Degree is a Java-based geographic server which supports WMS, WFS, WCS as well as other standards.

The existence and availability of open standards has enabled the tools mentioned above to be developed with adherence to internationally accepted approaches for online geospatial information exchange. This eliminates the need for resources in producing custom formats or APIs. As a result, these tools can communicate with any other tools which support standards, whether they are proprietary or open source, in a fairly transparent fashion.

1.5 Integration

There are numerous methods in which the combination of open standards and open source software can communicate. Two examples are shown in the following discussion, first within a given codebase and second as part of a distributed service-oriented architecture (SOA).

1.5.1 With FOSS4G Software

Using MapServer's Python MapScript, a developer can utilize the features built into the MapServer C API. MySQL supports spatial extensions via the OGC SFS specification. This means that spatial data fields in a MySQL database can be tested for various spatial predicates, as well as basic input and output/display. The following simple example shows how, using WKT, MapServer can re-project a coordinate from a MySQL database:

```
#!/usr/bin/python

import MySQLdb
import mapscript

projInObj = mapscript.projectionObj("init=epsg:4326")
projOutObj = mapscript.projectionObj("init=epsg:26918")
db = MySQLdb.connect(host="localhost",user="foo", passwd="bar",
db="mydb")

cursor = db.cursor()
cursor.execute("SELECT AsText(geo) FROM locations")
result = cursor.fetchall()

for record in result:

shape = mapscript.shapeObj.fromWKT(record[1])
shape.project(projInObj, projOutObj)
print shape.toWKT()
```

The example could have easily been implemented using MapScript in Perl or PHP, or any development library which is aware of WKT. Similarly, a PostGIS-enabled PostgreSQL database could have been serving the spatial data. This emphasizes how adherence to standards can make such data manipulations much easier to develop than would be the case otherwise.

1.5.2 As Components

Open source software can also leverage open standards as part of a larger information infrastructure. Consider, for example, a Web mapping application that integrates datasets through Web services. The client's connection code consists of a single approach to interact with any WMS or WFS server. The same approach is used for each layer requested. This results in leaner software codebases, given the abstraction and uniformity that open standards provide. That is, no matter what software is used, standards facilitate the client-server interactions. Figure 1.1 displays, at a very high level, this concept linking various GIS software (open source in shaded boxes, commercial in clear boxes). For example, if the MapServer WMS were changed to an Intergraph WMS server package, this would cause no disruption in the operation of the infrastructure because Intergraph and MapServer both support the WMS specification.

Hence, open standards ensure that open source projects can interoperate with each other, as well as with commercial packages, resulting again in "loosely coupled" infrastructures, based on a service-oriented architecture approach.

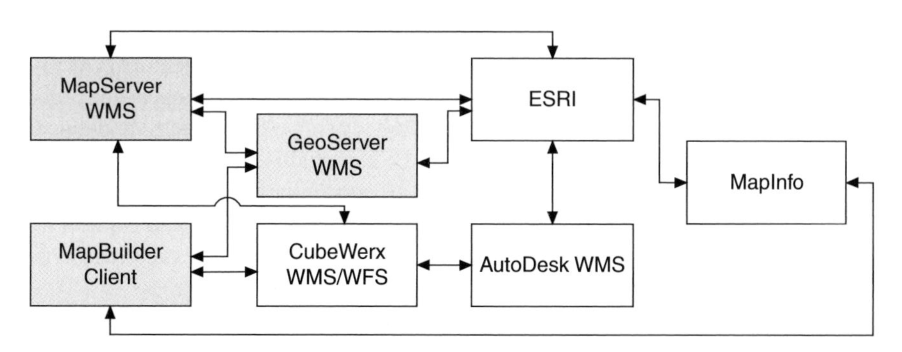

Fig. 1.1 Conceptual intregration of open source software as building blocks for a broader SDI

1.5.3 Example

Consider a scenario that uses MapServer to generate a WMS of earthquake information. The first step in "OGC-enabling" these data is to fetch and reformat the text records into an interoperable, self-describing format. Keep in mind that if this Web service existed, this step would not be required. The initial process is to establish the data format within the data located at http://neic.usgs.gov/neis/finger/quake.asc.

As these data have a geographic location as well as attribute information relative to a series of points on the earth's surface, a GML approach may be used. The primary step in creating a GML document is to create a GML application schema. This application schema defines the data types, structures and objects in W3C XML schema language. Because GML represents an enabling framework, which itself leverages XML schema, a domain expert can easily construct their information

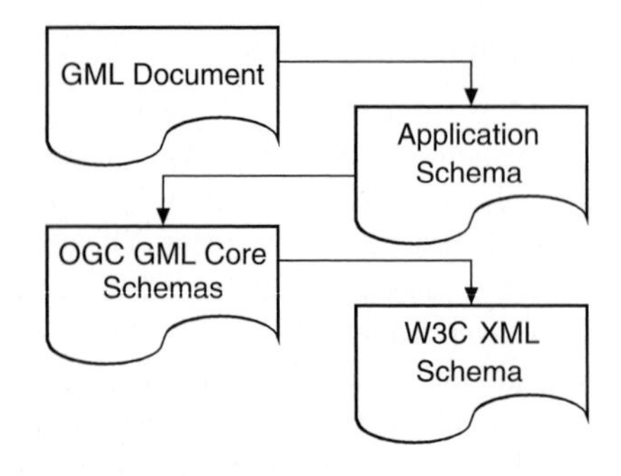

Fig. 1.2 Leveraging standards as information building blocks

model in a standards-based fashion. This reduces the level of effort required to define common nomenclatures and structures where others have already defined them. Figure 1.2 illustrates the building block effect of a GML document.

When creating the application schema, the GML core schemas possess many predefined blocks which can be reused. This saves time and effort by eliminating the requirement for redefining common blocks and structures, as well as (and more importantly) providing an output information model in a form that common tools can process and interpret.

As illustrated in Fig. 1.3, objects defined within the gml:AbstractFeatureType region indicate those inherited from the GML core model. The gml: namespace indicates reuse of an existing definition from the GML schemas. The objects are defined in the local application schema as specific to the National Earthquake Information System (NEIS) data model. A simple scripting process outputs a GML document as input for the WMS. A UMN MapServer installation is then configured to connect to the GML data. Source code, schema and configuration files for this example can be found at http://www.kralidis.ca/gis/eqmapping/.

Once this process is in place, the WMS can run stand alone and unsupervised as a Web service with a self-updating process to gather latest updates from the NEIS data site. As a result, any WMS-aware client application (Web-based or desktop) can interact with the NEIS data source for visualization, data extraction and/or analysis.

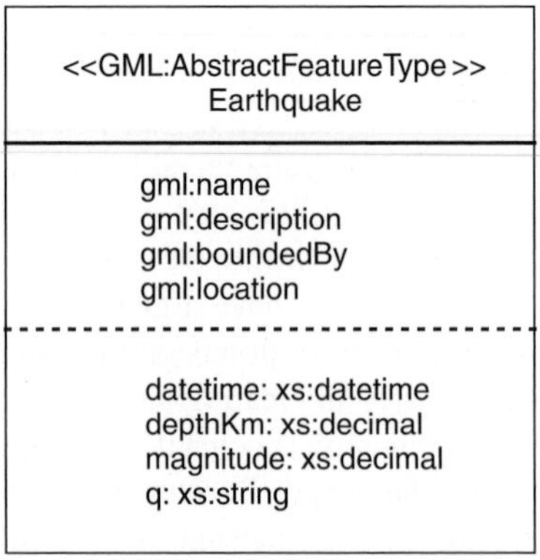

<<GML:AbstractFeatureType>>
Earthquake

gml:name
gml:description
gml:boundedBy
gml:location

datetime: xs:datetime
depthKm: xs:decimal
magnitude: xs:decimal
q: xs:string

Fig. 1.3 Schema design view of earthquake data GML model

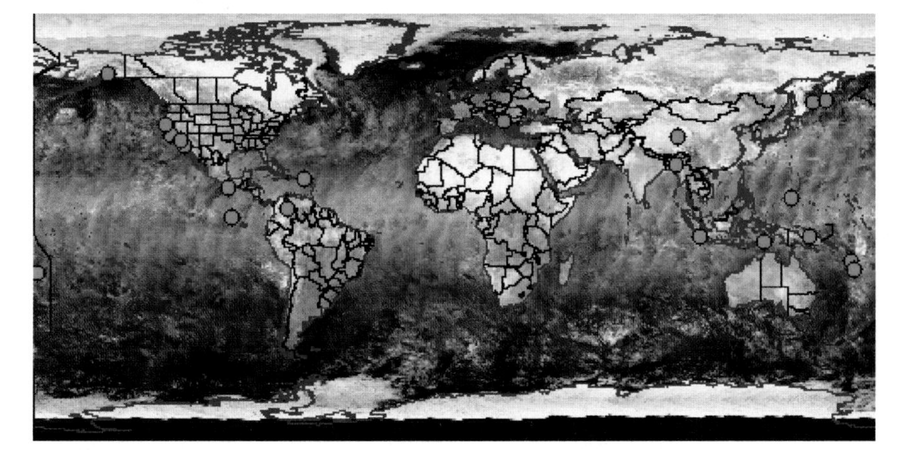

Fig. 1.4 Distributed data visualization: integrating envisat and NEIS from two different servers dynamically via WMS

A webpage could then automatically integrate the NEIS WMS and basemap data, which triggers two requests, producing the virtual Web map as shown in Fig. 1.4.

The WMS GetMap request embedded in the webpage to visualize the NEIS layer is written as follows:

```
http://geo.example.org/wms?
SERVICE=WMS
&VERSION=1.1.1
&REQUEST=GetMap
&SRS=EPSG%3A4326
&BBOX=-180.0000000309986,-112.5000000193741,180.0000000309986,
  112.5000000193741
&WIDTH=560
&HEIGHT=350
&LAYERS=neis
&STYLES=
&FORMAT=image%2Fpng
&BGCOLOR=0xFFFFFF
&TRANSPARENT=TRUE
&EXCEPTIONS=application%2Fvnd.ogc.se_inimage
```

To integrate the Envirsat layer as a backdrop to visualize NEIS, a similar WMS GetMap request would be invoked, changing only the LAYERS parameter and the location of the WMS serving the Envisat data.

This request is a valid WMS GetMap request connection, which means it can apply to any valid WMS server, given the correct URL location and content information. The advantage of using a standards-based API in this case is that it (or the logic used to generate it) may be reused and applied to an infinite number of applications.

The WMS specification, as a Web service, has also reduced barriers to integration for non-traditional spatial data users. For example, the Finance Maps Web site offers free analysis and information on global investing with a focus on personal finance (Good Content Network 2004). With the help of OGC standards and a WMS, this Web site now provides map displays of global financial information as a means of integrating with current financial marketing conditions and forecasts. Specialized resources are not required to develop simple references to OGC-based Web services thereby enabling ease of integration for applications.

With spatial data available via open standards, application developers can now integrate these data with far less effort than previous approaches allowed.

1.6 Future Developments

The OGC OWS Common Specification streamlines the often-used and defined area of OGC specifications. For example, WMS, WFS and WCS all provide basic metadata when returning a GetCapabilties document, all of which have subtle differences. OWS Common provides a unified framework for these specifications to use common constructs when defining metadata elements. This will result in leaner OGC specifications, which refer to OWS Common as opposed to attempting to define the same information without use of this standard. This subsequently means less code development for software packages, especially when they support numerous OGC specifications. This will also allow for more rapid implementation of new OGC specifications into software packages.

As suggested earlier, the popularity and rapid adoption of de facto standards such as GeoRSS will continue to offer options for information to be published using community-defined formats, at the same time as making their way into more formal standards bodies and processes.

1.7 Conclusion

The Internet has enabled networks of data and information to be exchanged across the Earth in a revolutionary fashion, the likes of which have never been seen before. Information communities, subject matter experts, scientists, researchers, and the general public can now communicate more efficiently regardless of physical location of content or networks. Open standards provide a basic common framework for information to be integrated in a consistent fashion across these networks. Organizations such as W3C, ISO, and OGC provide essential standards which lessen the complexity of software development and allow for independent and disparate tools, technologies, data and information to interact with one another more transparently than ever before.

Open source software can satisfy the growing requirement for interoperability by leveraging open standards, in addition to the benefits of open source software overall, thus contributing to SDI technology and exchange of spatial data for informed, location-enhanced, decision making.

References

Adaptive Path (2005) Ajax: A new approach to web applications, <http://adaptivepath.com/publications/essays/archives/000385.php>, 05 May 2007.

Aslesen L (1998) The protection and availability of digital cartographic information and products. In Taylor DRF (ed) Policy issues in modern cartography. Pergamon Press Inc., Oxford.

Australian and New Zealand Land Information Council (2007) Australian and New Zealand land information council website, <http://www.anzlic.org.au/>, 08 August 2007.

Begg C, Connolly T (1998) Database systems: A practical approach to design, implementation and management, 2nd edn, Addison-Wesley, New York.

Canadian Geospatial Data Infrastructure Architecture Working Group (2001) CGDI architecture description, version 1.0, < http://www.geoconnections.org/publications/tvip/Vision_E/CGDI_Vision_final_E.html>, 24 September 2003.

Community Mapbuilder (2007) Mapbuilder website, <http://www.communitymapbuilder.org/>, 14 May 2007.

Environmental Systems Research Institute (1998) Shapefile specification, <http://www.esri.com/library/whitepapers/pdfs/shapefile.pdf>, 03 December 2004.

Federal Geographic Data Committee (2004) Federal Geographic Data Committee website, <http://www.fgdc.gov/>, 03 December 2004.

Flanagan D (2006) JavaScript: The definitive guide, O'Reilly and Associates, Sebastopol, CA.

Gardels K (1999) The open GIS approach to distributed geodata and geoprocessing, < http://www.ncgia.ucsb.edu/conf/SANTA_FE_CD-ROM/sf_papers/gardels_kenn/ogismodl.html>, 23 May 2003.

GeoConnections (2007) GeoConnections website, <http://www.geoconnections.org/>, 23 January 2007.

GeoJSON (2007) GeoJSON website, <http://www.geojson.org/>, 15 May 2007.

GeoRSS (2007) GeoRSS website, <http://www.georss.org/>, 21 January 2007.

GeoServer (2007) GeoServer homepage, <http://www.geoserver.org/>, 14 May 2007.

GeoTools (2007) GeoTools homepage, < http://geotools.codehaus.org/>, 07 August 2007.

Global Spatial Data Infrastructure Association (2001) Developing spatial data infrastructures: The SDI cookbook, Version 1.1, <http://www.gsdi.org/pubs/cookbook/>, 12 October 2003.

Good Content Network (2004) Finance maps personal finance website, <http://www.financemaps.com/>, 02 December 2004.

Google (2007) KML webpage, <http://code.google.com/apis/kml/documentation/>, 15 May 2007.

Groot R, McLaughlin J (2000) Geospatial data infrastructure: Concepts, cases and good practice. Oxford University Press, Oxford.

Guerrero I (2004) Interoperability – Why it makes good business sense for our industry, <http://directionsmag.com/article.php?article_id=520&trv=1>, 25 February 2004.

Hartman R (1997) Focus on GIS component software. Onward Press.

Harvey F (2000) The social construction of geographical information systems, In International Journal of Geographical Information Science, Volume 14.

Infrastructure for Spatial Information in Europe (2007) Infrastructure for Spatial Information in Europe website, <http://www.ec-gis.org/inspire/>, 08 August 2007.

International Organization for Standardization (2007a) International organization for standardization webpage, <http://www.iso.org/>, 10 April 2007.

International Organization for Standardization (2007b) International organization for standardization technical committee 211 webpage, <http://www.isotc211.org/Outreach/Overview/Overview.htm>, 10 April 2007.

International Organization for Standardization (2007c) International organization for standardization technical committee 211 webpage, <http://www.isotc211.org/Outreach/Overview/Factsheet_19115.pdf>, 10 April 2007.

International Organization for Standardization (2007d) International organization for standardization technical committee 211 webpage, <http://www.isotc211.org/Outreach/Overview/Factsheet_19139.pdf>, 10 April 2007.

International Organization for Standardization (2007e) International organization for standardization technical committee 211 webpage, <http://www.isotc211.org/Agreements/Agreement_OGC.pdf>, 10 April 2007.

JSON (2007) JSON website, <http://www.json.org/>, 15 May 2007.

National Aeronautics and Space Administration (2004) National Aeronautics and Space Administration website, <http://www.nasa.gov/>, 24 July 2004.

Open Geospatial Consortium (2006) GeoRSS White Paper, <http://www.opengeospatial.org/pt/06-050r3>, 10 February 2007.

Open Geospatial Consortium (2007a) OGC implementing products, <http://www.opengeospatial.org/resource/products>, 18 February 2007.

Open Geospatial Consortium (2007b) SensorWeb enablement overview, <http://www.opengis.org/functional/?page=swe>, 18 February 2007.

Open Geospatial Consortium (2007c) Digital rights management, <http://portal.opengeospatial.org/files/?artifact_id=17802>, 18 February 2007.

Open Source Geospatial Foundation (2006a) Tiling standard, <http://wiki.osgeo.org/index.php/TilingStandard>, 15 May 2007.

Open Source Geospatial Foundation (2006b) Tile map service specification, <http://wiki.osgeo.org/index.php/Tile_Map_Service_Specification>, 15 May 2007.

Open Source Geospatial Foundation (2006c) WMS tiling client recommendation, <http://wiki.osgeo.org/index.php/WMS_Tiling_Client_Recommendation>, 15 May 2007.

Ramsey P (2004) A survey of OGC deployment, < http://web.archive.org/web/20050831193022/http://digitalearth.org/story/2004/12/1/15658/1000>, 02 December 2004.

Refractions Research (2005) PostGIS website, <http://postgis.refractions.net/>, 13 May 2007.

Regents of the University of Minnesota (2001) Codebase mapwms.c history logs, <http://trac.osgeo.org/mapserver/changeset/332/>, 15 May 2007.

Regents of the University of Minnesota (2007) MapServer website, <http://mapserver.gis.umn.edu/>, 15 May 2007.

United States National Research Council (1999) Distributed geolibraries: Summary of a workshop, <http://www.nap.edu/html/geolibraries/>, 18 November 2001.

World Wide Web Consortium (2007) World wide web consortium website, <http://www.w3.org/, 01 June 2007.

Yeung AKW, Hall GB (2007) Spatial database systems: Design, implementation and project management. Springer, New York.

Chapter 2
Free Software and Open Source Business Models

Arnulf Christl

Abstract This chapter discusses the nature of free and open source software from the perspective of business models that can be used to operate within this emerging industry. The focus in the chapter is on free and open source software in general rather than on the specific geospatial domain, however this area of activity is used in several places for reference and by example. The chapter also examines the closed and proprietary complements of the free and open source world and contrasts the rationale and behaviours of software developers and users within both market places. While it is not the intention to set one or the other markets up as a straw man, it is clear from the discussion that the free and open source alternatives have a number of advantages in terms of developing high quality outputs that are responsive to end user needs while embodying the principles of innovation and advancing knowledge.

2.1 Introduction

One of the first questions often asked of business people working in the free software industry is: "How do you make money if you give away the software for free?" Typically, this question is followed by an exclamation mark and a quizzical look in the eyes of the questioner that is meant to demonstrate that this is actually a very good question. In fact, this question is not hard to answer, although it is hard to convey a complete understanding of the answer to those working outside of the free software industry. In part, this is due to the perception of software as a product that is bought and owned, which is, itself, a concept that is learned by individuals at a very young age during basic socialization within a free market economy. However, there is more to software than just owning a copy of a product that can be legally used. Specifically, all software products have associated installation, customization, maintenance, training and upgrade needs, which vary considerably from product

Arnulf Christl
WhereGroup GmbH & Co. KG, Siemensstr. 8, 53121 Bonn, Germany; Open Source Geospatial Foundation, Beaverton, OR, USA, e-mail: arnulf@osgeo.org, arnulf.christl@wheregroup.com

G.B. Hall, M.G. Leahy (eds.), *Open Source Approaches in Spatial Data Handling.*
Advances in Geographic Information Science 2,

to product. This aspect of the computer software industry is substantially more complex that the selling of usage license fees, as consumers never actually buy a software product when they purchase it. Rather, they purchase a set of temporary usage permissions that must be agreed to before the product can be legally installed and used.

There are several reasons why it is important not to be impatient when people ask the above question. The most important is that often the starting point of understanding the free software industry is often ignorance, and this can only be overcome through education, removing ignorance and, even more importantly, removing false interpretations and twisted logic that result from applying the wrong concepts to the development and role of free software in a free market economy. The foremost of these false interpretations is to think that software is the same as any other physical good that can be traded. This chapter takes the view that this is not the case and develops an argument that supports the reasons why.

There are strong economic arguments that make corporations want consumers to believe that software is just another good. Terms like "commercial, off-the-shelf software" convey the character of a product. There are many more examples how to "personalize" software that is originally just a copy of a unique set of computer readable instructions. In some cases a unique (license) key is added to make a copy an individual, personalized entity. Sometimes software is even tied to hardware and whenever the hardware breaks or has to be exchanged the software will not run any more. All of this adds up to a fair amount of work that must be done to propagate and sustain the notion of software being just another good.

In addition to the Free Software industry, there is a growing community of developers who work within the Open Source community. Their predilection to produce and distribute source code for other developers to enhance has, in some cases, led to a neglect of understanding the basic underlying concepts that are required to make a living from using, deploying, developing, consulting, training and supporting software. Hence, this chapter deviates somewhat from the spatial data handling context of the remainder of this book and, instead, focuses on Free Software business models in general. Despite this, there is a clear spatial aspect of Free and Open Source Software (FOSS) that is embodied in the globalization mantra "Think Global, Act Local". While FOSS is being developed actively all over the world, there still is a need for direct, personal contact at the local level.

2.1.1 Disclaimer

This chapter describes FOSS as a legal concept on the one hand and as a development philosophy on the other hand. To show the advantages that FOSS offers to users, developers and businesses it is contrasted to proprietary models which have dominated the public view on the software industry for at least the past twenty years. Especially in the late 1990s proprietary software companies started actively to antagonize the FOSS movement because it endangered the proprietary business model. This fight was and is supported by heavily funded campaigns and seemingly independent lobby organizations. The resultant spread of Fear, Uncertainty and Doubt (FUD) of the FOSS movement emerged partly out of ignorance of the advantages offered by alternative business, licensing and governance models, but it was also purposefully engineered to discredit FOSS and its development communities.

It is important to keep in mind in the following discussion that any licensing model of software is no proof of quality, regardless of whether it is free and open or closed and proprietary. To achieve good results with FOSS business models, it is not enough simply to apply an Open Source (OS) software license. Furthermore, not all software that is licensed as FOSS is automatically any better than proprietary software alternatives.

In February 2006 several leading FOSS communities bundled efforts to create the Open Source Geospatial Foundation (OSGeo) to support and build the highest-quality open source geospatial software tools possible. To achieve this, software projects go through an incubation process within this organization prior to their release and licensing (http://www.osgeo.org/incubator/process/process.html). Hence, there is a stamp of quality assurance that follows the same proven development and governance models that this chapter endorses. While this is only one approach to a development process that has business viability twinned with quality assurance and standards adherence, it is an approach that has achieved some success to date.

2.1.2 Business or Ethics and Both

There is a tradition to consider economics and ethics separately as though they are incompatible concepts. Unfortunately many commercial activities are unethical but completely legal, and sometimes the impression can arise that unethical but legal business is the most lucrative and expedient path toward commercial success. The seeming tensions between these concepts should be considered and a course charted that is ethical and legal as well as being able to foster business development. In this context, altruism is ethically high-valued and can therefore appear as remote from business activities. Because OS and especially the Free Software movement seem at first sight to be altruistic in intent, they are also regarded as being remote from viable business models. However, this reasoning is wrong, first because FOSS development does not have to be motivated by altruism, and second because business and ethics do not have to mutually exclude each other.

This chapter identifies a common ground between the two seemingly opposed perspectives and shows how to build a business with FOSS activities. The question of whether an enterprise is ethical or behaves in an ethical way can probably not be answered from a general viewpoint, and surely not in this short introduction to business around the FOSS model. Thus, the chapter tries not to judge business models but only to show which of them are viable in the long run.

2.2 Definitions of Software, Free Software and Open Source

This section provides an outline of the concepts described by the terms software, Open Source and Free Software. The discussion is only cursory, however it is necessary to go into some depth of understanding of the nature of software because it is fundamentally different from the physical products that consumers normally purchase. This discussion is necessary to gain an understanding of the business ideas behind associated FOSS models. As noted earlier, the discussion is general in nature and does not make a specific case for FOSS over other forms of software.

For practical reasons it is nowadays appropriate to abbreviate the activities of interest as FOSS. However, to explain the concepts behind FOSS it is helpful to consider the two aspects of FOSS separately. One is best described by the term OS as it focuses on the development methodology. The other is better described by the term Free Software (FS) as it regards the licensing model and legal background. These basic concepts are explained in the next sections.

2.2.1 Software and Hardware

To understand the nature of software it is helpful to discuss the ways in which it differs from hardware. There are more differences in the prefixes "soft" and "hard" than similarities in the suffix "ware".

Computer software cannot be executed (used) without hardware and vice versa. Hence, it is fairly common not to see the two terms as separate concepts. When the first computers were built the software was built alongside of them and there was no such thing as an IT industry that thrived on producing only software. The first computers consisted of tons of wires, bulbs and switches and were manufactured by scientists and electronic engineers (such as the Zuse Z11, which sold only 5 computers in 1956 – see http://www.epemag.com/zuse/part7c.htm). Every computer in that early era had its very own individual set of instructions.

Only much later, with the emergence of the personal computer in the early 1980s, was it possible to use one copy of software separately by one producer on the hardware of another (IBM was one of the major facilitators of this process). This was a sort of revolution that brought software its freedom of individual existence. However, this freedom did not last long because the newly emerged independent software allowed for a highly scalable business model. The task of implementing a software product only has to be funded once, but the results can be duplicated (copied) infinitely often at practically no additional cost as this does not require the same resources as the original product (essentially the creativity of the developer(s)).

To reduce the highly volatile and easily duplicable character of software it has become common practice to "bundle" software with hardware, such that if the hardware changes the software may stop running. The reason for this is not technical, as the above described independence of software from hardware has reached a very high level. Rather, the reason is that the license basically prevents changing the configuration of the bundle. This confusion between hardware and software causes all kinds of trouble in understanding FOSS concepts that are focused explicitly only on software. However, hardware is a material concept, to which physical laws apply. Software, on the other hand, is immaterial and therefore physical laws do not apply.

A few examples that help to demonstrate this are shown in Table 2.1. Following the logic of this table, it is apparent that business models focusing on the availability of physical goods cannot apply naturally to software. Hence, the definition of what constitutes software in physical terms becomes difficult. To lessen the potential for confusion, software vendors have invented a new term, the "software product". This is essentially an attempt to make software resemble a physical good. Once software comes into existence it is naturally free of the limitations of physical availability (regardless of whether it is titled a "product" or not) since its supply is virtually endless at zero or a very small marginal cost. After the software has been implemented, all natural limits to its distribution are reduced to network connectivity costs or the trivial costs of the physical media that it is published on.

Many problems result from regarding software as just another good, or a form of solid ware, that is sold in boxes "off the shelf". As software has no physical representation and is made up of volatile bits and bytes, obviously other laws must apply. Perhaps it would be more appropriate to stop using the term "software" altogether and instead use something more appropriate like "nonware" or "technolyrics". This latter term nicely conveys two basic concepts inherent to software in that "techno" refers to the technicality of the (non)-"thing" at hand, while "lyrics" address the creative human component that the software is comprised of.

Table 2.1 Basic differences between hardware and software

Hardware	Software
If hardware (or any physical good) is sold the supplier suffers a loss of that good that can be compensated by payment.	If a copy of a software "product" is sold or given away, the original copy still exists. Hence, the supplier suffers no physical loss of the product because it is only a copy.
If hardware breaks or ceases to function beyond easy repair, it becomes useless.	Software cannot break in the same sense. A data carrier can be scratched (for example a CD) but the original of the software is not affected.
Hardware cannot be duplicated. Every copy needs the same amount of raw material and energy as any other. Copies of complex hardware will always be imperfect.	Software can be duplicated completely. Each successful copy of a software product is an identical reproduction of the original (the "raw material" is the source code, it does not run out).
Hardware can wear out, rust, or decay, and will eventually break and cease to function.	Software does not wear down, rust, decay or break. It may fall out of use, but it never loses its basic functionality.

2.2.2 The Early Days of Software Development

In the early days of computing, the original and natural way to solve computing problems was scientific collaboration. Every new software developed was based on prior knowledge and added some new aspect to it. At this point in time there was no need to use the term "Free Software", since there was nothing to set this way of creating software apart from anything else.

Original (free) software development was only slowly displaced by proprietary thinking in the mid-1970s when it started to become economically viable to produce software that was independent of hardware. One document that shows this emerging line of thought incidentally is a letter sent by Bill Gates on February 3rd, 1976 (Gates, 1956). The content of this letter is void of the above inherent difference of software and hardware and confuses fairness with business models. In fact, the letter is not directed to the professionals of that time but to a group Gates describes as hobbyists. It turned out later that these "hobbyists" were in fact the "users" around whom Gates would create his business empire. In that same letter Gates asks: "Who cares if the people who worked on it get paid?" The obvious answer is that in the existing market economy model people need to take care of themselves and have to work out a way to get paid before they start to work. This is the basic difference between what makes a hobbyist a professional and it has nothing to do with fairness. The process of turning software into "property" was gradual and came hand-in-hand with a differentiation of hobbyists into users and developers.

2.2.3 The Open Source Development Model

Source code is the part of software that is human readable. It contains the information that is required to enable the software to run. All changes to software (removing errors, adding functionality, enhancements etc.) are done in the source code. Before software can be executed by a machine the source code has to be transformed (compiled) into binary or object code. This process is typically irreversible. Once software has been compiled it cannot be changed. Most end-user software nowadays comes in a compiled binary format and is shipped without the source code. Without the source code, software cannot be modified, repaired or enhanced and any control over what the software does is significantly restricted.

Open Source (OS) and the associated logo are trademarks of the Open Source Initiative (OSI) (http://www.opesource.org). Any software claiming to be OS has to abide by the terms and conditions defined by the OSI. These terms and conditions specify that the source code of open software must be published fully and without restrictions. Anybody can take OS software and look into its inner workings, change, improve and give away any number of copies to anyone else. Software is not seen as a secret hidden inside a black box, but as a living resource from which more and better software can be produced. OS licenses ensure that this concept is based on a sound legal foundation. The legal background coincides almost completely with that of the FS definition which is described later in this chapter.

The other factors that make up a good OS project concern governance, communication and organization. This chapter only touches upon the topics that make the OS development model successful. A comprehensive set of instructions on how to produce OS has been compiled by Karl Fogel at http://producingoss.org, and this resource is an absolute prerequisite for anybody who wants to understand the implications of opensourcing (as a verb) their software. A lot of trouble can be eliminated if more people would educate themselves about these basic principles, including those who feel the need to argue for or against OS models. Two key factors emerge from Fogel's guidelines on OS development, namely "publish early and release often".

2.2.4 Publish Early

OS software is often started as a solution to a concrete problem. The sooner the solution is published the better for the project in the long run. This is crucial for all good OS software projects as it allows others to join the process at an early stage. If the solution is good then others may also be able to use it and start to contribute. Contributions can come in many different forms including actual software development as well as funding, documentation, testing, even publicity or recommendations of the software to other developers and users. The more people who pick up on the idea, the more testing, enhancements and development the solution will experience. Good communication right from the start is one of the crucial aspects of any OS project and sets it apart from closed development.

It can also happen that there already is a solution or other groups who intend to do the same things or have already progressed farther. In those cases it can save a lot of time and money not reinventing the wheel by joining efforts on a common project. It can also be an incentive to try and be better than the competing project, as diversity is good for any natural development. Moreover, the traditional grassroots process has a more spontaneous nature than a planned project that is organized from top to bottom with traditional power structures.

2.2.5 Release Often

Rapid prototyping and agile computing paradigms are followed by many projects. Publicly accessible code repositories allow developers and users alike to pick up the latest changes on a daily basis and keep up to date. Changes are documented in the repositories and distributed through special mailing lists so that anybody who is interested can follow edits in the code base very closely. Software in general is never finished and always buggy, and OS projects are no exception to this rule. However, there is a higher likelihood that bugs in OS software will be resolved quickly and corrections to the software are typically added as patches that can be applied as regularly as desired. However, a proper software release is more than just the latest code from the repositories. It should be a tested, approved and as stable as is reasonably possible for public release. Release cycles should not be dictated by commercial considerations or marketing dates. Still, as a general rule, releases should come in a regular fashion and procrastination of release dates should be avoided as much as possible.

Basically, there are two different ways to determine release dates. One release type focuses on the availability of a defined set of new functionality that has been added to the software, or whenever it becomes awkward to stay up to date by applying a large number of patches at once. The other type defines an interval within which the next release is to be published. Depending on the amount of changes, the version number will then be incremented. For a collection of patches only the third version digit (the patch number) is increased. For changes or additions in functionality the second version digit is raised and the third set back to zero. Only deep functional changes, a complete re-engineering of the software and broken backward compatibility will cause the first version number to be changed.

Many of these development basics have also been picked up in slightly modified ways by large proprietary vendors who release patches and intermediary versions in regular intervals, sometimes even on a daily basis.

2.2.6 Free Software Licensing Model

The other factor that makes the OS approach to software development successful is the FS licensing model. It was developed in the early 1980s by Richard M. Stallman (http://www.gnu.org) in response to the growing possessiveness of businesses with respect to software developed by their employees. Stallman's mission was to use legal means to protect all intellectual work with a license that made sure that it could not be enshrined as individual property.

Stallman used a straightforward legal approach to achieve this. First, the work was put under copyright protection. This is the standard way to protect software or any creative work under both national and international copyright conventions (such as the international Berne Convention of September 9th, 1886 and its numerous revisions through to the present). Legal measures can be taken to protect the interests of the copyright holder. After having protected the software by a copyright covenant, distribution terms were added that made sure that everybody could use, copy, give away and modify the software – given that the changes are published using the same license. Thus, all software that ever appeared under this license would remain protected and tied to the same distribution terms.

Actually the requirement that all changes be tied to the same license was the most confrontational aspect of Stallman's license approach because it effectively excluded all proprietary business models that "protect" code by hiding it from the rest of the world. The most prominent example of this kind of license is the General

Public License (GPL) of the GNU project (http://www.gnu.org). This kind of license is sometimes called "viral" with reference to its potential to "infect" non-free software. In the same mindset it could also be seen as a vaccination for software to protect it from ever becoming proprietary. Another term frequently used to describe the effect of this license is "copyleft". It is symbolized as a reversed c in a full circle like the copyright symbol, but mirrored. However, unlike the copyright symbol it has no legal meaning.

2.2.7 From Free Software to Open Source and Back

One of the pivotal developments in the conceptualization of FOSS was the publication in 2001 of Eric of Raymond's book, "The Cathedral and the Bazaar" (http://catb.org/∼esr/writings/cathedral-bazaar/). In this book he summarized differences between a distributed, rapid development methodology and the traditional waterfall model. Specifically, he attributed the prior approach to OS and the latter to proprietary development. At that time it actually appeared as if the bazaar was the metaphor of the OS approach and that the only limitation for more commercial uptake was the software freedom mindset introduced by Stallman.

In another article, Raymond started an ongoing debate about the ambiguity of the term "Free" in "Free Software", which in English language can be associated with "gratis" as easily as with "freedom" (http://catb.org/∼esr/open-source.html). In order not to discourage commercial businesses he proposed to start using the term "Open Source" instead of "Free Software". This caused friction within the FOSS community finally leading to a schism that resulted in the creation of the OSI and long debates with Free Software Foundation (FSF) activists.

Bruce Perens, an early Debian GNU/Linux lead, was the primary author of the OS definition. This can be used to determine whether a software license can be considered to be OS or not, according to the following principles:

1. Free Redistribution: the software can be freely given away or sold. (This was intended to expand sharing and use of the software on a legal basis.)
2. Source Code: the source code must either be included or freely obtainable. (Without source code, making changes or modifications can be impossible.)
3. Derived Works: redistribution of modifications must be allowed. (To allow legal sharing and to permit new features or repairs.)
4. Integrity of the Author's Source Code: licenses may require that modifications are redistributed only as patches.
5. No Discrimination Against Persons or Groups: no one can be locked out.
6. No Discrimination Against Fields of Endeavor: commercial users cannot be excluded.
7. Distribution of License: The rights attached to the program must apply to all to whom the program is redistributed without the need for execution of an additional license by those parties.

8. License Must Not Be Specific to a Product: the program cannot be licensed only as part of a larger distribution.
9. License Must Not Restrict Other Software: the license cannot insist that any other software it is distributed with must also be open source.
10. License Must Be Technology-Neutral: no click-wrap licenses or other medium-specific ways of accepting the license must be required.

The above terms of the definition of OS software are almost completely in line with the definition maintained by the Free Software Foundation (http://www.fsf.org/licensing/essays/free-sw.html). Hence, for all practical purposes Free and Open Source Software go together well. What makes the real difference between FOSS and proprietary software nowadays is the distribution terms, not the development model. Thus, the concept of FOSS is better conveyed by the term Free Software and its associated mindset.

2.2.8 Legalizing IT

The OSI lists 65 licenses (http://www.opensource.org/licenses/December 2007) that comply with the above terms and conditions and can thus legitimately use the OS trademark. These licenses cover a broad range from the restrictive Copyleft to more permissive ones that also allow the code to be used and distributed under proprietary terms and conditions. All of these licenses have gone through the OSI License Approval process (http://www.opensource.org/approval) and have a documented history of being applied to numerous OS software projects. Some of the licenses have also been approved in court in different national jurisdictions. This comprehensive repository is one of the references for the end user who wants to find out whether the software of choice is protected by a known and proven license.

Parallel to the OSI website, the FSF website lists 64 licenses (http://www.fsf.org/licensing/licenses/#SoftwareLicenses December 2007), many of which are also on the OSI list. The FSF adds a short description to each license identifying 30 that are compatible with the GNU GPL and 34 that qualify as Free Software licenses but conflict with the properties of the GNU GLP. These licenses have gone through the FSF Free Software Licensing and Compliance Lab which was informally established in the year 1992 and was formalized in 2001. From an end user's perspective all licenses are legally applicable. For developers, the FSF recommends to use only GNU compatible licenses as this allows developers to include and reference all software that is protected by the GNU GPL license (roughly three quarters of all OS projects).

These FOSS licenses are legal constructs that can be enforced in court. Hence, there is no difference between proprietary and FOSS licenses with respect to the legal power that can be exerted to enforce licensees to abide by the terms of the license. It is usually much harder for an individual (regardless of whether the individual is a user, developer or representing a business) to find independent information about proprietary license schemes and their applicability than for FOSS licenses.

Proprietary licenses are usually not developed through a consensus process and can be worded individually. Proprietary licenses can be and are changed at any time by the issuer. The above listed references to approved FOSS licenses have higher longevity, and are also a lot more stable and reliable.

The underlying general differences between proprietary and FOSS licenses becomes apparent when comparing license texts, especially with respect to the distribution terms. All proprietary licenses contain a lot of detail on what users are not allowed to do, whereas FOSS licenses explicitly grant users a variety of rights. These rights include the permission to copy, redistribute, re-engineer and modify the software or any parts of it, and, most importantly, to give away the software to be used by anybody else.

The restrictions that proprietary licenses have, in combination with the virtual and easily copyable nature of software in general, lead to a high potential of criminalization of the end users. FOSS licenses work the other way round by explicitly defining what users are allowed to do, especially to copy, change, modify and redistribute the source code and object code or any part of it. Hence, using FOSS licenses makes life easier for users and reduces the risk of breaching license terms and breaking the law.

2.3 Life Cycle Versus Development Cycle

Figure 2.1 shows a conventional software life cycle model compared to the OS development cycle model. The left hand side of the diagram shows some examples of how the life cycle model is implemented in a proprietary environment and reference to the right side shows how it compares to the OS development cycle.

Proprietary software vendors typically impose artificial life cycles on software to force users to upgrade to newer versions. In fact, some licensing schemes have an end date after which the software may not be used any more even if it would work appropriately. Users then have no other choice other than to upgrade to a newer version because the old version is no longer supported. Hence, it falls out of use and the "death" suggested by the term "life cycle" is caused by marketing mechanisms designed to keep the machine running.

Upgrading to new versions of software may require the payment of additional fees, often on a regular basis, by signing a software maintenance contract. Such a contract is intended to protect the user's initial investment in acquiring the usage license for a package. Without maintenance, the permission to use the software will end after a period that is defined by the license issuer. This effectively means that the initial cost of acquiring a usage license is lost. Hence, from this perspective the cost is not an investment but a one time expenditure. This issue is especially important in the spatial domain where there still is a heavy dependency between software and data in the sense that many vendors implement proprietary (closed) data formats that can only be read by the software they were created in.

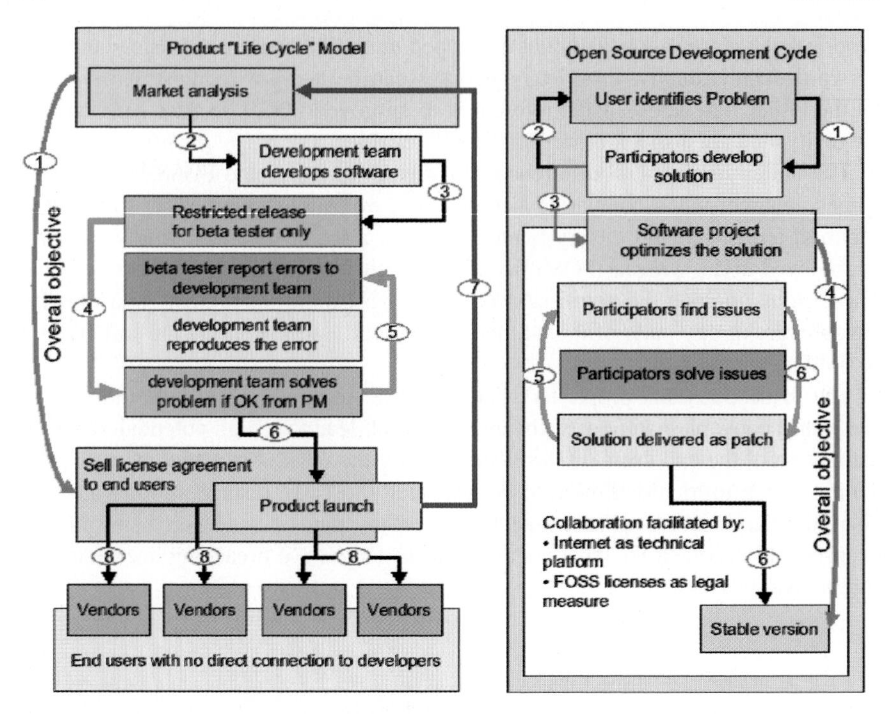

Fig. 2.1 Development cycles
(Source: Arnulf Christl 2007, http://www.mapbender.org/presentations/AGIT/modified)

Proprietary software users usually have no mechanism to influence the direction of development or new functionality that may be added to a package. This decision is taken in the best interest of the software vendor who may or may not take into account what the users want. Major releases often require migration of data, adjustment to new interfaces, and potentially migration of dependent software like applications running on top of a database. In contrast, OS environments allow for a faster iteration of development cycles because there is no profit-oriented marketing overhead that has different interests in where development goes.

The overall objective of most proprietary software vendors is to maximize profit (1) by selling as many license agreements to end users (8) as possible. The initial product design will be thoroughly influenced by a preceding market analysis that tries to predict the expected user behavior for the coming years.

After designing the "product" it goes into the planning and implementation phases (2). This is where core development takes place and it is not different to the FOSS development cycle. Test cycles start when the first beta version is ready for release. It can be distributed to selected beta testers who, in some cases, may have to pay for the privilege to be the first to be able to test the software. The hardware industry, as an example, has a vital interest in testing their components with the newest operating systems as soon as possible to be able to present compatible hardware when the software is released to the market.

The beta testing cycle iterates for a predefined time that is scheduled according to the marketing plan. Errors found during beta testing (4) may or may not be resolved (5). This is up to the product management team that controls the overall speed and path of development. Sometimes enhancements will not be implemented immediately but deferred for the next version to make long term maintenance contracts more attractive. There must be convincing arguments to make the user want to upgrade to the newest version.

At some point in time the stable version will be released. Usually the release date does not coincide with the software development being "finished". There are several reasons for this, the first being that software is always a work in progress and never really is finished. Another reason is that the introduction of new software or a major release change is a delicate undertaking. If it fails, the whole carefully planned schedule is in danger of failing. Depending on the type of software, releases will usually coincide with a major industry event, trade fairs or the period immediately prior to Christmas. The time to release proprietary software to the market is when end users are most receptive to chose it. There are plenty of examples where the schedule did not work out as planned, and this is sometimes reflected in the product name as a year number.

After launching the product vendors and resellers start to distribute the software (8) and the whole process iterates (7) as long as the market analysis promises high enough revenues for overall profit (1).

The overall objective (4) of the OS development cycle shown on the right hand side of Fig. 2.1 generally focuses on producing stable software that solves a problem. The cycle starts because someone has a problem (1) that can be solved by using software (2). If the solution is good it might attract more developers (3) and spawn a new project.

In the OS realm the terms "developer" and "end user" are much less clear cut than they are in a general sense. Every developer most of the time is also a user of other software, be it an email client, Web browser, operating system, or database. This is true regardless of whether the developer works in an open and free environment or in a proprietary environment. Users are the most important factor when testing software for usability and market acceptance. OS environments encourage communication between users and developers, greatly reducing time-to-market for new features and bug fixes. Because there is no explicit vendor role that divides developers from users they can interact more easily and participate in a collaborative work process. This justifies using the much more appropriate term "participant" for software developers and users alike.

A new software project will only come into existence and develop if enough people participate. Many OS software projects get born, wither and die because nobody uses them (6) and nobody has a reason to continue to develop them (5). Software stays in use as long as this cycle continues to iterate. In a proprietary environment this iteration can be kept going artificially for some time before the software actually dies. On the other hand, this iteration can also be stopped at will and at any time by the proprietary owner and copyright holder of the source code. This often happens in the course of software acquisitions by proprietary enterprises who then discontinue development. In contrast, it is less likely for this to happen in the FOSS software development model because it is always possible to pick up the latest available version and continue development. This is also the reason why OS software is not threatened by corporate takeover or merger, and another reason for the difficulties that large enterprises with an exclusive proprietary business model are currently faced with.

2.4 FOSS Governance: Taking Back Control

The core of the proprietary business model is to maintain full control of software development and explicitly to limit the users' rights on the software. This means that users have no control over how software is developed, not to mention when and which new functionality is added or removed.

As noted earlier, a specific characteristic of the geospatial realm is the high dependency of software on data. Without geodata, geoprocessing software is rendered useless. A lot of spatial analysis needs to be able to process data across different domains and thus potentially across different software architectures. The need to be able to access datasets regardless of the hosting software environment is so strong that it is becoming more and more difficult for proprietary vendors to create typical vendor-lock-in situations by implementing closed formats.

FOSS governance models open up the development process by a group that takes decisions in an open and transparent way. There are many different governance models depending on the history and size of the project, the intended target audience and user groups. Small projects are often initiated by a single developer who is automatically in control of the code repository. As new developers join the project the initiator might choose to open up governance or keep it centralized. Quite a few projects can flourish perfectly well under a centralized model if the head is a good leader. One such project is the Geospatial Data Abstraction Library (GDAL http://www.gdal.org/) started by Frank Warmerdam in the late 1990s (see Chap. 5 of this text). As GDAL became more and more important to many other projects and businesses, its governance was opened and a project steering committee was created (currently with five members). This ensures that development of the project is not controlled by the interests of one individual.

Other projects such as Mapbender (http://www.mapbender.org/), initially started in 2001 as proprietary software that was implemented and owned by a single company, namely CCGIS, which was the forerunner of the Where Group (http://www.wheregroup.com). After releasing it as FS under the GNU GPL license in 2003 the development process gradually changed. A formal project steering committee was then created to allow enterprises and users who made heavy use of the software to join this group. When Mapbender became a formal OSGeo project it went through the OSGeo Incubation Process (http://community.osgeo.org/incubator/process/process.html), which included a thorough revision and enhancement to the governance model. Currently, members representing companies, users and individual developers form the Mapbender steering committee (http://www.mapbender.org/index.php/PSC).

Independent of the governance model chosen for a FOSS project, the license terms always allow a project to be developed in a different direction than that planned by the group initially in control. This process can be detrimental to project development and can lead to the dilution of effort. One such example is the Java Unified Mapping Platform (JUMP) that is currently maintained by several different communities including:

- http://openjump.org/
- http://jump-project.org/
- http://www.vividsolutions.com/jump/
- http://www.saig.es/en/kosmo.php
- http://deegree.sourceforge.net/src/demos.html
- http://jump-pilot.sourceforge.net/.

To avoid this kind of dilution of resources, successful projects must put a lot of effort into meeting the needs of users, developers and businesses and to run and operate the project with an open and transparent governance style.

2.4.1 Decisions in a Voting Processes

A good example for an open, collaborative governance model is the MapServer project (see Chap. 4 of this text). It started as a research project that was later joined by a private company, DM-Solutions Group (http://www.dmsolutions.ca/), as well as individual contractors. The project is very successful and has grown a large community with very different needs. To be able to address better the broad spectrum of requirements, a formal entity was needed. The first official Request for Change (RFC 1, http://mapserver.gis.umn.edu/development/rfc/ms-rfc-1/) describes how this new MapServer Technical Steering Committee determines membership, and makes decisions on MapServer technical issues.

The voting system is very common throughout OS projects. In brief, the technical team votes on proposals on the developer mailing list server. Proposals are available for review for at least two days, and a single veto is sufficient to delay progress though ultimately a majority of members can pass a proposal. The following steps describe this process in detail:

1. Proposals are written up and submitted on the mapserver-dev mailing list for discussion and voting, by any interested party, not just committee members.
2. Proposals need to be available for review for at least two business days before a final decision can be made.
3. Respondents may vote "+1" to indicate support for the proposal and a willingness to support implementation.
4. Respondents may vote "−1" to veto a proposal, but must provide clear reasoning and alternate approaches to resolving the problem within the two days.
5. A vote of −0 indicates mild disagreement, but has no effect. A 0 indicates no opinion. A +0 indicate mild support, but has no effect.
6. Anyone may comment on proposals on the list, but only votes from members of the Technical Steering Committee will be counted.

7. A proposal will be accepted if it receives +2 (including the proposer) and no vetos (−1).
8. If a proposal is vetoed, and it cannot be revised to satisfy all parties, then it can be resubmitted for an override vote in which a majority of all eligible voters indicating +1 is sufficient to pass it. Note that this is a majority of all committee members, not just those who actively vote.
9. Upon completion of discussion and voting, the proposer should announce whether they are proceeding (proposal accepted) or are withdrawing their proposal (vetoed).
10. The Chair gets a vote.
11. The Chair is responsible for keeping track of who is a member of the Technical Steering Committee.
12. Addition and removal of members from the committee, as well as selection of a Chair should be handled as a proposal to the committee.
13. The Chair adjudicates in cases of disputes about voting.

Interaction and communication is achieved by using public mailing lists (http://lists.umn.edu/archives/mapserver-dev.html), forums, wikis, IRC chats, conferences and meetings. Especially, public mailing list archives contain a wealth of meta information about a project's past development, current health, and might even allow some predictions about its future. The main reason why many developers are not yet consciously involved in OS development methodologies is because they work in environments in which the underlying proprietary business model relies on software being treated like a scarcity, and its inner workings as a secret. Any public or open voting system would simply not work with this business model.

2.4.2 Limitations of Retail FOSS Business Models

Both OS and, especially, FS are frequently associated with a general anti-business ideology. From the limited perspective of a proprietary software vendor this might seem true because selling individual license usage agreements for software that is also available for download at essentially no cost at all is difficult, if not impossible.

In the early days of OS, business was recognized mainly as a gift economy (see Raymond, 2001), which did not scale well and lacked a solid foundation for enterprise businesses. Most information published around OS business models still focuses on the traditional approach in which an enterprise wants to make money by producing one particular ("their own") software. This only works for software that can be distributed in a retail style as a commodity allowing for high scalability and keeping the price per unit low. Each unit is an enabling technology that through its use will generate an added value that is higher than the individual cost, or else there would be no reason to purchase the software. The overall revenue generated by leveraging this enabling technology is a multiple of initial licensing costs, as Bruce Perens shows with the example of Microsoft:

> Microsoft is a tool-maker, and the effect of the tool-maker on the economy is tiny next to the economic effect of all of the people who are enabled by the maker's tools. The secondary economic effect caused by all of the people and businesses who use an enabling technology is greater than the primary economic effect of the dollars paid for that technology. And of course the same is true for Open Source software. Perens, Feb. 16 2005 (http://perens.com/Articles/Economic.html)

The results of this approach are recipes to make a profit by slightly bending Free and Open definitions or by combining them with proprietary models from the outset. Koenig (2004) identifies seven strategies to generate business and leverage OS:

- The Optimization Strategy
- The Dual License Strategy
- The Consulting Strategy
- The Subscription Strategy
- The Patronage Strategy
- The Hosted Strategy
- The Embedded Strategy.

These business strategies usually involve mixing proprietary and FOSS-based models, or are a vehicle to enhance business generated by selling hardware. They are hybrid business models derived from the needs of large enterprises and, for the most part, do not apply to small scale businesses or to highly specialized areas like the geospatial domain.

These niches are better addressed by small and medium-sized enterprise (SME) approaches and especially by pure service providers. From this perspective FOSS allows the adherence to much more attractive business models because it is a lot easier to provide services for a software without restrictions concerning its distribution than when distribution is restricted in some way. This distinction is probably the most important between proprietary (retail) and FOSS (services) business models. It also shows that the focus does not lie on one specific software but on the potential to use any software from the FOSS world to build specific solutions. However, this grassroots-oriented approach does not scale well for large enterprises.

Another reason for the general trend to associate FS with an anti-commercial attitude is that IT companies still rely heavily on proprietary business models and do not recognize the advantages offered by FOSS. Large companies with a long tradition of selling software usage licenses have an infrastructure that has been optimized mainly to market products. By releasing their software under a FOSS license they would destroy this source of income and would go out of business quickly. Therefore measures have to be taken to ensure that the proprietary business model will survive long enough to build a more reliable business model that can also take the specialties of FOSS into account.

Nonetheless some large enterprises have also started to embrace OS and use it to enhance their business portfolio. In many cases these efforts still ignore the fundamental aspects of FS licensing schemes and governance models and only try to make a profit from using OS (for example the GNU Linux operating system) instead of proprietary third party software (like the Microsoft Windows operating system).

Depending on the targeted market, services around software make up an increasingly larger share of potential revenue than income generated through collecting usage licenses fees. In the consumer market, scalability is so high that individual license costs can be kept at a relatively low level. Specialized software with smaller markets cause higher costs per installation making it more difficult to keep license fees on a low enough level not to scare away potential customers. Especially in the geospatial realm, additionally, a lot of professional expertise is required to get the software to run with the data and interact with other components. This is true regardless of the underlying software licensing scheme, and in this respect there are no inherent advantages for the application of either proprietary or FOSS licenses.

A further characteristic of spatial data is their longevity. Spatial data are persistent and do not easily migrate from one system to another. Often, the act of migrating data is a lot more expensive in terms of work load than what initial licensing costs plus long term maintenance costs of the underlying software add up to. The migration from proprietary to FOSS-based environments opens up a source of income focused on FOSS. The subsequent professional operation of the infrastructure is another traditional source of income that also applies for proprietary software.

2.4.3 Hybrid Proprietary/FOSS Business Models

Anybody can use FS to make a profit by using it for commercial activities or by providing services. As noted earlier, there are no legal complications in using FOSS licenses in a business or commercial context (http://www.ibm.com/developerworks/opensource/newto/#8). This also includes software that has been protected by Copyleft licenses (for example the GNU GPL). The only limitation is that it is not allowed to bundle and resell modified versions without opening the corresponding modified sources.

Large enterprises with a long history of selling software usage licenses have a difficulty in adopting FOSS business models because they operate complex software product design, marketing and distribution networks. The technical need has been rendered obsolete by the emergence of the Internet as a distribution platform. In the professional environment, the need for costly marketing campaigns is reduced due to most information being available ubiquitously via the Internet, newsgroups, domain communities and so on. In this regard the fact that FOSS licensing models have created new fields for commercial activities that are not focused on selling boxes but on providing services is a distinct advantage.

International Business Machines (IBM, http://www.ibm.com) has recognized this development and over the past years has developed a hybrid business model. IBM offers services for many OS software components and installs a growing share of hardware with the GNU Linux operating system, Apache webserver, MySQL database and PHP scripting language (LAMP) stack and distributes the Firefox Web browser, OpenOffice.org suite and so on with their retail hardware. The release of the integrated development environment Eclipse (http://www.eclipse.org/) as FS further opens up the developer market. IBM can make revenue within this market as a proprietary software vendor with its specialized development extensions. Besides this proprietary business model, IBM is also the largest software patent holder worldwide.

The business model of Oracle Corporation (http://www.oracle.com/) focuses on selling proprietary software usage licenses and on dominating the market by acquisition of other companies. Nonetheless Oracle has started to develop an OS plan with the argument that this reduces costs and increases stability for its customers. The corresponding text on the companies Web page reads:

> Today, many customers are using Oracle together with open source technologies in mission-critical environments and are reaping the benefits of lower costs, easier manageability, higher availability, and reliability along with performance and scalability advantages. (http://www.oracle.com/technologies/open-source/index.html December 2007).

Applying these positive attributes to the competing OS database system PostgreSQL (http://www.postgresql.org/) shows how difficult it can be only to adopt a partial OS strategy.

Another Oracle strategy to address the growing OS economy is to address directly the developer and Internet Service Provider (ISP) markets. The release of the gratis version of Oracle (XE, express edition) is intended as an incentive for developers (OS and proprietary alike) to implement software that runs exclusively on top of the Oracle database. This will enlarge the overall market share for Oracle and bind new customers through vendor-lock-in. This kind of commercial activity does not leverage or support FOSS, neither does it add much to the revenue generated by selling licenses but it does help to grow market share.

The geographic information system (GIS) company ESRI Inc. also operates on a proprietary business model and generates revenue by selling software usage licenses and maintenance contracts. ESRI also supports the use of OS software components but almost exclusively from other domains and in limited ways, for example as an alternative to proprietary operating systems, software development environments or the Web server. Geospatial OS software, on the other hand, is in direct competition with ESRI's software and associated business model. Using the OS database PostgreSQL instead of the proprietary equivalent from the Oracle Corporation will reduce overall costs for the user. But at the same time it also opens up the potential of PostGIS (http://www.postgis.org) to ESRI users which might lead to fewer users favoring the proprietary equivalent offered by ESRI.

ESRI is also a shareholder of the company 52N (http://52north.org/) that has been founded to promote the conception, development and application of FOSS4G software for research, education, training and practical use. The governance model of the software released by 52N is still evolving and currently comprises a dual licensing scheme with the GNU GPL and a proprietary license. This allows members of the consortium to package the software together with proprietary components.

One outstanding example of a large proprietary enterprise in the geospatial realm that has gone partially OS with its own software is Autodesk Inc. It has released a completely rewritten version of its formerly proprietary software MapGuide as OS (http://www.osgeo.org/mapguide) and donated the code to OSGeo to underline its commitment (see Chap. 7 of this text). In this context OSGeo serves as an independent legal entity to which community members can contribute code, funding and other resources, secure in the knowledge that their contributions will be maintained for public benefit.

Another OS initiative by Autodesk is the Feature Data Object (FDO) project (http://fdo.osgeo.org/). This is an application programming interface (API) for manipulating, defining and analyzing geospatial information for a variety of spatial data sources, where each provider typically supports a particular data format or data store. Here one of the strategic difficulties is caused by the strong ownership restrictions that Autodesk enforces for its proprietary computer assisted design (CAD) data formats and the growing need of the geospatial community to access, read and write CAD formats.

Marketing hybrid business models is very difficult because these models have to cover the gap between FOSS and proprietary sources of income. Most hybrid models only last for a transition period and have to undergo constant changes.

2.4.4 FOSS Services Business Models

Commercial activities around FOSS have a service character that is also applicable to most proprietary software. However, it is a lot easier to perform services using FOSS because license terms do not restrict distribution of the software. Additionally businesses explicitly focusing on leveraging FOSS will also help to further and support FOSS software development and the surrounding community. This is why hybrid models over time tend to erode the proprietary components. The main areas of FOSS business in the geospatial realm comprise:

- Education and Training
- Consultation
- Installation and Maintenance and Support
- Core development, implementation of new features.

Each of these services can be performed exclusively or in a value chain that covers all aspects from the inception of a software project through to long-term maintenance. FOSS can add value to all of these services instead of taking away potential sources of income as might appear at first sight. To complete this chapter, the following sections describe the detail of each service type and map it to real world examples in the geospatial FOSS domain.

2.5 Education and Training

Education and training are traditional ways to make a living within the knowledge domain. The basic idea with this form of activity does not really differ when training users of either proprietary or FOSS licensed software. The business model behind training is straightforward. Participants pay a direct cost for the time that the trainer spends teaching the software.

Using proprietary software in training may raise costs due to additional license fees and reduce the potential income for the trainer and both the company providing the training and also the institute receiving the training. However, since training is a precondition to efficient use of software most proprietary vendors have a special educational licensing program that reduces these costs. Starting at elementary school level and going up to university and research institutes, proprietary software vendors grant usage licenses at reduced fees or for no cost at all.

The reasoning behind giving away software licenses to students for free is straightforward. Specifically, it allows them to become educated using a specific brand of software and whenever trainees complete their education and move into a position where they can influence decisions for or against a software package they will naturally tend to recommend the software they know best. This also justifies why proprietary vendors spend considerable effort and costs on the preparation of tutorials and related educational material.

This motivation is much less pronounced in the FOSS world as competition with other projects has less of an economic advantage. The drawback of this relaxation of the competitive edge is that considerably less effort (if any at all) goes into the preparation of training materials and coursework. One major critique of OS software has always been the lack of good documentation and training materials. Some voice the opinion that this is being done on purpose and is a covered up business model (Fotescu 2007). However, the reason is more likely to be embodied in the competing demands that are put on participants within the OS community.

The obvious solution to this deficiency is to apply the same collaborative principles that make OS work in general produce coursework, tutorials and documentation. However, unfortunately, even in education, proprietary mindsets have recently started to spread.

2.5.1 Creating Course Materials

During development of a software typically only basic technical documentation is created. Some development environments and languages render this kind of documentation automatically. In all cases it is created by experts (developers) and intended to be used by experts (other developers).

Universities are highly scalable resources of education, but there is little incentive for professors to organize their students to produce thorough documentation and tutorials for any software that they create or contribute to in their coursework as long as a well known set is already available from a proprietary vendor. In the meantime, professional trainers create documentation for software packages, but they will not do it for free. They either have to charge more money for individual classes to recover the cost of creating the documentation, or to have a budget to cover startup costs.

One way to create excellent course materials is to do it collaboratively. The documentation produced by the developers is the basis on which professional trainers can produce course material. This course material is published along with the software package and can be used, extended and enhanced for free. University courses are ideal places for thorough peer review, and enhancements can go back into the course material. During university courses, it is often required to create homework which can be integrated into the larger curriculum. Practical exercises and (scientifically) verified solutions round up the effort that collectively has the potential to create an excellent source of tried and tested documentation and practical exercises.

Professional trainers can supplement their practical exercises to improve the training materials with real-world aspects that would also help to put some reality into university-based curricula. Yet again, all of this is only possible if all documents created during these work flows are published under license with no restrictions in the distribution terms. Two of the better known examples are the Creative Commons license family and the GNU Free Documentation License (GNU FDL).

There are good examples in the geospatial world that demonstrate how this approach can work. The software project GRASS (http://www.osgeo.org/grass) has excellent educational resources due to its long term involvement with university education (see Chap. 9 of this text). Other projects that are more deeply embedded in the software architecture of many FOSS4G projects like PostGIS need more technical and less scholarly documentation and training materials. In this case the documentation is created and is maintained by the project itself (http://www.postgis.org/documentation/).

2.5.2 Audiences and Participants

There are three distinct audiences for training each with individual business opportunities, namely:

- Software developers, experts and trainers
- Professional users
- University and scholarly education
- Proprietary think trap.

The first group can make a living out of providing training, consultation, customization and development of software. Individuals in this audience are usually capable of processing all required information around a software package without the need for a trainer. Instead, their background knowledge and experience gives them the capability to acquire the knowledge about a new software package through self-training. Good software documentation will help them get started more easily. If the documentation is poor or locked away, the software will not be used or distributed as widely as it might be. The competitive advantage for the software project is that developers and experts act as multipliers.

The second group spends money. It is the audience that actually needs information and background to software packages to help them solve their problems. They might not have the expertise that developers have or lack the time to learn through use and experimentation. They are prepared to invest in learning how to use the software to solve explicit problems that they need to solve efficiently. If this comprises learning how to use the software within a short time frame through professional trainers, then they will adopt this approach. This type of individual or group training requires a trainer of a high level of proficiency, and justifies higher prices.

University and scholarly audiences comprise students and professors. Software itself is becoming an important asset for conveying information and helps students learn. Additionally, competence in software helps students to qualify for the job market upon graduation. However, professors need to invest time in software developments in order to keep up to date, otherwise the chances are high that they will not help disseminate new ideas and may paradoxically become obstructions to the process of knowledge growth and dissemination.

This kind of learning is slow but thorough. There is time to enhance details, ask critical questions beyond the daily chores of coding, and do research. During courses, written work and, increasingly, multi-media content are created. All of this work can become a valuable part of an overall software project if it is protected by a free and open license and published online. Technologies to do this are Wikis, collaborative books, content management systems, document management systems and so on.

The fourth audience is often explicitly targeted by proprietary vendors with special license rates. Vendors perceive this subvention as a long term investment. This is because, as previously mentioned, students who have learned how to use one package of software will tend to use that package when they start to work in the professional market.

2.5.3 Providing Teaching

More than anything, the costs of a trainer vary by location. A local trainer in Brazil will be able to offer much better conditions than somebody coming to Brazil from Germany, the United States or Australia. Simply not having to travel half way around the world will reduce the overall costs of training. Among the other benefits of operating locally, bills do not need to be issued across borders, taxation issues are easier, and people can meet more easily to prepare the teaching materials. Many local points of presence are required to offer teaching in native languages, hence this is a good opportunity for local service providers.

There are topics and software packages where the expertise initially lies with one or only a few individuals. This is the unique selling proposition (USP) that these individuals can exploit until the innovation has disseminated. The expertise is naturally first adopted by prior scholars. This competitive advantage allows an individual or a company to be one step ahead of the others, however it is lost or at least reduced during training. Thus, it is necessary to stay ahead of the rest of the world, which is only possible by continually increasing capabilities and expertise.

Exploiting a single temporary advantage (leading know-how) is not a good foundation for a long-term business model. Rather, it is more sustainable in the long-term to maintain technical advantage through ongoing innovation and ingenuity, which is one of the driving factors in FOSS models.

Teaching or training can be as diverse as the requirements of the trainees. Depending on the type of or approach to training that is adopted, it can be useful either to implement a hands-on learning environment, lecture using static presentations, or work with a combination of both. In general, there are three distinct ways to educate in the use of software:

1. *Introductory information.* This type of teaching helps to orientate trainees, and provides a high level overview of relevant topics. It is usually conducted by individuals who are used to giving presentations focusing on one topic. Conferences are an ideal framework for this kind of teaching.
2. *Dedicated software courses*: This approach usually teaches trainees how to use one software package, or it has a distinct focus on one topic. This form of training is usually named after the software or topic. For example, in the case of "MapServer training" people will expect to learn how to use the MapServer software package. In the "Multi Band Satellite Imagery Analysis" course, people will expect to learn how to do just that, potentially with several different software packages.
3. *Problem-driven workshops.* This approach typically addresses one particular problem and ideally results in an ad hoc solution that precisely addresses the problem at hand. This type of training blends into consultation.

2.6 Consultation

Complex geospatial information, analysis and data acquisition frequently involve the use of several different software packages and the integration of different data sources. No single software package solves all problems. Increasingly, the term geospatial is taken to mean a way of looking at things with a perspective that takes location into account. Within this domain different software packages can, more often than not, be used to produce almost identical results with the same data. Sometimes it will be advantageous to use one package over the other. However, in many cases the evaluation process for selecting which package to use can be quite complex.

This is where professional consultation is important. This form of employment is possibly one of the best paid hourly rate jobs within the IT industry. The reason is that educated decisions have to be taken within a restricted time period and need to take a variety of factors into account. The education and training required to be able to make the right decision need the investment of a lot of time and money, as a wrong decision can have significant consequences for many years. Hence, good consultation can be expensive, yet it is attractive from a business perspective both to the client and the consultant given its potential long term implications.

Taking software into operation for a customer involves considering many parameters. This is not special to FOSS but to software in organizations in general. The amount of money required for proprietary usage licenses usually only makes up a at most a third of what is needed for the overall effort required to install and maintain complex software systems. The cumulative costs resulting from using inappropriate software and the losses caused by unproductive employees over time can reach much higher levels than the initial acquisition cost of a proprietary license.

Vendor-lock-in can bind users to a specific software package for a long time, and prevent it from being substituted with other software, even if it becomes apparent that the system in use causes problems. Further, migration from one software system to another often raises numerous problems (e.g., converting data, changing workflows). Thus, consultation should always focus on using software that implements open standards and can be used easily in conjunction with software from other sources (see Chap. 1). Another advantage of the FOSS approach is its transparency and the possibility to start using it without initial expense.

2.6.1 Installation, Maintenance and Support

Installation of desktop applications can be straightforward, but in some instances they can also be somewhat daunting, especially if the application is unpackaged or if the application does not comply with the IT-security policies of an existing environment. Depending on the type of software and system environment under consideration, business models based on performing installations can be very different. For example, for server software it may be necessary to compile special versions that fit neatly into the customer's existing environment.

The maintenance of operating environments can be performed by client staff (who need to be educated and trained), but increasingly this level of maintenance is also outsourced to service providers. The complexity of geospatial data and processing in part is due to the niche character of the software. There is little or no scalability potential in spatial data infrastructures as each focuses on specific domains of interest. The emerging spatial software stack of OSGeo can be used to build a foundation for geoprocessing capabilities but the interaction of the components first has to be designed and then implemented within the existing architecture. This level of implementation and its maintenance typically require a relatively high level of knowledge, and support contracts can help to consolidate an infrastructure as well as reduce the requirement for deep internal knowledge among clients.

Support contracts are more flexible than pure maintenance contracts in that they can be used to maintain the system or to enhance features or extend functionality. Much of what is typically provided as first level support can be obtained through mailing lists for free but this cannot be fixed in a formal contract. Second level support requires more knowledge of the application of the software in use and is requested by integrators or other service providers, who either provide first level support themselves, or are bound to provide guarantees which need to be covered by back-to-back agreements.

Third level support needs intricate knowledge of the inner workings of software and is often provided by the core development teams of software projects. This level of support can also help to leverage development of the software in the direction that is needed by the customer.

The business model is, again, straightforward. The client reports a problem that is solved by the service provider. Agreements to limit excessive support hours can vary from contract to contract as there are no general rules as to how these work. In many cases clients buy a support quota that is worked off by the service provider on request.

2.6.2 Software Development

Software development is the core of what is commonly associated with FOSS approaches. This aspect of work is probably the most difficult activity to make a living from. This is at least true for software developed in the FOSS4G realm, as it is a niche activity rather than a commodity. The scalable commodity type of spatial applications has been picked up amazingly fast by large Internet companies leaving little room for new software, but opening up a whole new secondary market of application mashups that are based on the new geospatial commodities.

There are different development methodologies that vary depending on what language the software is implemented in, and what kind of solution is addressed. Many FOSS projects start as any grassroots movement, namely as modest contributions of coding that solve specific problems of relatively limited scope. If such software solves a problem and people use it, then it has the potential to improve and grow. Once it has reached a certain level of acceptance, or is being used in professional contexts, chances are high that it will be further improved and consolidated. Over time, the software and the people involved in implementing it will mature, and development methods will accordingly professionalize. If the software is a complex package or involves many dependencies, sooner or later it will be necessary to organize it in order for the software to be sustainable in the long-run. The job of organizing and coordinating good development is usually not visible on the outside (the user's perspective) of a software package. Thus, it is hard to argue for money to support the development effort.

Generic code that can be used for many purposes and does not address only one need or solve one problem is the most difficult aspect of development to find funding for. Usually, this kind of development has to be cross-financed from implementing features that are often transparent to end users. The only way to argue for a generic approach is to be completely transparent about the required development, and to explain why the resources invested in developing generic software will pay off in the long run. This is also a task that usually cannot be performed by a single developer. Rather, it needs project-level organization and, at best, an independent contact point for the project. In the FOSS4G world, this task can be taken on by a project or technical steering committee, that consists of a group of developers and power users who share the responsibility to "run" the project.

The independent contact point can also be a professional who problem solves for a fee (for example by making a support contract). In many cases the solution provider will be part of the development group of that software, but this is not a requirement. The ubiquitous availability and unconditional accessibility of FOSS allows anybody to enhance, repair and improve it, for money or out of any other motivation they may have (personal gratification in solving a hard-to-solve problem, peer accolades etc.).

In the FOSS community, depending on the license and the development contract, the implementer should or even must give the enhancement back to the rest of the world unencumbered by proprietary licenses. This is also in the interest of the customer because it enhances the chance that his/her extension will become part of the main software, and thus be available along with updated versions of the main software without additional implementation effort.

If the source code is locked away as-is in a proprietary environment, there is a natural monopoly that accrues to people who have access to the source code, hence those who can will contribute to its development. This inevitably results in less diversity and it ignores the potential of achieving the highest quality by opening up the code to the scrutiny of the maximum number of peer reviewers. Thus, the chance to implement longevity and robustness is foresaken.

As a client it is easy to adopt the quickest and (from a short-term perspective) cheapest solution. This will invariably be a hack. If the problem can be solved with a hack and afterwards the implementation falls out of use (e.g., one-time converter software), then this is a perfectly appropriate way to operate. Even then it makes sense to publish the snippets and fragments of code that led to that solution, enabling others to profit from the work that has already been done. To stay with the example of the converter, it will help people to convert their data into the new format more easily, which might produce follow-ups on the software around the new format. One such example noted earlier is the GDAL project which has a long history of collaborative development and gives generic access to a broad range of different formats.

The project hack appears to be the better solution for the short-term from the viewpoint of the customer and end-user because it is cheaper and addresses the immediate problem directly. However, in most cases it will pay off in the long-term to implement a generic solution because of the above arguments. This insight often reveals itself to the customer only after he/she has encountered a few painfully hard brick walls of frustration. It can be difficult to lend somebody who has gone through this frustration a helping hand and still insist on FOSS being the better concept. Thus, it is very important first to educate the customer of the implications of a specific strategy, and then provide good consultation evaluating the advantages of each implementation option.

2.7 Conclusion

This chapter has provided a review of free and open source business models and contrasted them against their counterparts from the closed and proprietary world of software development. The chapter started by providing definitions of software, free software and open source as a backdrop to the subsequent discussion. A differentiation was made between hardware and software as this is fundamental to the differing business models that can be adopted. The OS development model was then discussed and several of its underlying premises were noted. This was followed by consideration of the Free Software licensing model, and the relationship between the two was discussed. The body of the chapter examined the life cycle and development cycle of free and open source software and proprietary software, and concluded with a comprehensive review of the business orientation, rationale and limitations of the free and open source software community. In this discussion the education and training markets and the role of consultant services were given special consideration.

References

Christl A (2007) <http://www.mapbender.org/presentations/AGIT/modified>

Fotescu R (2007) The sorry state of the Open Source today, <http://beranger.org/feature/sorryfeature.php> 28 July 2008

Gates W (1976) Open letter to hobbyists, Homebrew Computer Club Newsletter, 2, 1, p.2

Koenig J (2004) Open Source Business Strategies <http://kp.cospa-project.org/retrieve/1537/opensourcebusinessmodels.pdf> 29 July 2008

Perens, Brue (2005) <http://perens.com/Articles/Economic.html>

Raymond E (2001) The Cathedral and the Bazaar: musings on Linux and Open Source by an accidental revolutionary, O'Reilly Media Inc., Sebastapol, Ca.

Chapter 3
Communities of Practice and the Business of Open Source Web Mapping

David McIhagga

Abstract The growth in adoption of open source Web mapping technologies in the past five years has been nothing short of spectacular. Although a great deal of focus is given to the technologies themselves, the genesis of their growth has been in the social and economic trends that have driven technology development and adoption. This chapter focuses on the emerging open source Web mapping ecosystem and the many stakeholder groups that play roles within it. In particular, the chapter outlines the powerful relationships between businesses and partnerships between business and communities of practice around open source technologies. By understanding these relationships, technology users, contributors, corporations, and communities can identify their own roles within the ecosystem and from this establish strategies to achieve success.

3.1 Background – Why Open Source Web Mapping Technologies Maximize Value to End Users

Web mapping technologies have grown rapidly in their use around the world and nowhere has this growth in adoption been more extreme than in the adoption of Open Source (OS) solutions. The world's attention has focused, since Google acquired Keyhole Technologies on October 27th, 2004, on the battle of the titans between Microsoft and Google to see who will dominate the new world of mapping "mash-ups", with Google boasting up to the time of writing of over 10,000 applications built on top of Google Maps. Meanwhile, a small OS project originally developed at the University of Minnesota (see Chap. 4 of this book) is estimated to have an install base of over 100,000 worldwide. This leads to the questions of who is the real titan? and how and why did all of this happen?

David McIhagga
DM Solutions Group, Ottawa, ON, Canada, e-mail: dmcilhagga@dmsolutions.ca

G.B. Hall, M.G. Leahy (eds.), *Open Source Approaches in Spatial Data Handling*.
Advances in Geographic Information Science 2,

3.1.1 Web Mapping – The Ultimate Solutions-Driven Technology

The answers to these questions lie not in the technologies that have evolved, but in how they are used. Examination of a cross-section of the highest quality Web mapping applications available reveals several things in common. Paradoxically, the most important commonality is the plurality of differences, since the value of Web mapping is not so much in the power of the capabilities they provide, but in how these capabilities can be accessed, presented and interfaced by radically different user communities.

OS Web mapping technologies, unlike their proprietary counterparts, have overwhelmingly been developed and extended on the basis of the needs of specific application requirements and user communities, both commercial and non-commercial. This has occurred because core Web mapping technologies are but tools that serve the needs and demands of the continually increasing number of communities that seek to incorporate spatial content into their Web application requirements.

Hence *value*, is not so much in the Web mapping technology itself, but in the application of this technology to meet a demonstrable community need. The net effect of this lop-sided weighting in value favouring solutions over technology has changed the dynamic of software development for Web mapping. Instead of value being in the generic software itself, the premium value is in the ability to advance the technology to meet high-value application needs. The result is the early stage development of an ecosystem that enables commercial and volunteer activities to answer these needs through the advancement of OS Web mapping technologies within the broader free and open source software for geospatial (FOSS4G) community.

3.1.2 Breaking away from GIS

The disruptive impact of the shift in value from intellectual property to the ability to advance technology has turned the traditional approach of the geographic information systems (GIS) industry on its head. Perhaps, the methodology and approach that worked so well to develop mass-produced, ever-growing, feature-rich GIS "products" is flawed in a world of solutions-driven Web applications. If solutions truly are driving the technology in Web mapping, then it is important to understand exactly how this mechanism works and to understand what the consequences are for developers and users alike.

Clearly, understanding the mechanics of this new paradigm for software development that already exists in the OS Web mapping universe will help software developers and users to take advantage of this changed and exponentially growing world. The result will be the more rapid emergence of a strong, and healthy OS Web mapping ecosystem.

3.2 The Open Source Web Mapping Ecosystem

The characteristics of Web mapping technologies represent some curious contradictions:

- Web mapping is and will likely remain a "niche" specialization, yet demand for the technology comes from virtually every definable industry.
- Core Web mapping technologies are rapidly becoming a commodity, yet the combination of skills required by an organization to deploy Web mapping are unique.

These characteristics define three of the central reasons why Web mapping technology has seen so much success as OS projects:

- A niche market with demand from literally a wildly differing world of industries does not provide the economies of scale needed to drive a successful proprietary product.
- A result of the unique combination of skills required to deploy Web mapping solutions is a significant weighting of value towards services versus products, and this is a strong characteristic of successful OS projects.
- Demands of specific vertical markets create commercial opportunities to deliver proprietary solutions that incorporate OS Web mapping technologies.

Although these characteristics favour the development of core OS Web mapping technologies, they also demand breadth in terms of skills, knowledge and organizational structures that are not conducive to having all of these activities take place within one or even a few organizations. Some examples of this diversity include:

- *Diversity of Skills* – including C/C++ software development, GIS analysis, cartography, Web application developers, and Web designers.
- *Diversity of Knowledge* – includes understanding of software development, geospatial information, Web mapping + domain knowledge from every unique community requiring Web mapping technologies.
- *Diversity of Organizational Structure* – includes all levels of Government, enterprise solutions, software as a service, and software providers, to name a few.

Developing an organization that can contain any one of these groups of skills, knowledge or organizational structures, is extremely difficult if not impossible. However, delivering any one solution around Web mapping requires the right combination of skills from each of these areas. In order to meet this demand, solutions are being developed through partnerships, and the emergence of a flexible and dynamic ecosystem around OS Web mapping technologies is serving to bind such partnerships into a common working environment.

The Web ecosystem is currently in its infancy. However, its characteristics are already quite clear. It is functional, growing, and maturing, and an interesting array of stakeholders who collectively form the ecosystem are beginning to emerge.

3.2.1 Stakeholders

The emerging ecosystem is a result of the coalescence of a series of changes to work, hobbies, and technologies as well as more broadly-based societal changes. These include, among others, the dominance of the Internet and the international connectivity it delivers, ever improving tools for collaboration, a cultural shift to using the Internet as a backbone for work and daily lives, and the ability for anyone to share and distribute information.

These societal and technological changes, when combined with the disruptive nature of the development of Web mapping technology described earlier, have created a real and thriving ecosystem around OS Web mapping technologies. Due to the complexity and diversity of skills, knowledge and organizational structures required to deliver Web mapping solutions, a set of four distinct stakeholder roles have emerged in the ecosystem. These include:

- *Caretakers*: Developers who build and support the core OS Web mapping technologies.
- *Professionals*: Power users and Web mapping experts who deploy solutions for others.
- *Specialists*: Domain experts who strive to incorporate spatial components for a specific industry, domain or market segment.
- *Consumers*: Stakeholders who use these technologies and provide some form of positive contribution within the ecosystem.

The stakeholders who fill these roles, represented by individuals and the organizations they operate within (if any), do not have a clear line of where they operate across the roles. They do, however, typically have a core commitment to at least one of the stakeholder roles.

Note that there are many beneficiaries of OS Web mapping technologies whose roles are not discussed in this chapter. Since this discussion is focused on the interactions and drivers of the OS Web mapping technology ecosystem, only stakeholders and their respective roles that contribute directly or indirectly to the advancement of these core technologies are included. These roles are described more fully in the following sections.

3.2.2 Caretakers

The caretakers within the OS Web mapping ecosystem consist of all individuals and organizations that directly contribute to the software development process or support others in the use of these technologies. Contributions come from across the spectrum including commercial entities, individuals, non-government organizations, Government departments and academia. Notably, the most successful OS Web mapping projects have at least one commercial entity making significant contributions to the project as caretakers.

The role of a caretaker is simple. It is to ensure that all other stakeholders can maximize the benefits of using the technologies being developed and supported by this group. Being a caretaker is not the exclusive domain of developers. It covers all aspects of the software development and maintenance lifecycle required for the production of high quality software. This includes testing, documentation, release management, and back-end technology support.

The drivers for acting within the ecosystem as a caretaker are as varied as the contributors to the Web ecosystem itself. For some, it is simply a hobby. For others it is a strategy for managing the risk of using OS technologies. For others it is a means for ensuring that a key component of a whole specialized domain is not only developed, but also maintained to ensure both high standards in its production and reproduction.

Most importantly, not all of these services are necessarily provided "for free". In fact, for some companies, the caretaker role is the most highly valued service available for their customers and partners. In general, the drivers are inevitably different for all caretakers. However, they do share a common goal and need for high quality Web mapping software that meets the needs of all other stakeholders.

3.2.3 Professionals

As discussed previously, the skill sets required for the delivery of high quality Web mapping solutions are highly diverse and specialized. As a consequence, a large and important group of stakeholders in this ecosystem serve the role of Web mapping experts, or power users, for the delivery of Web mapping solutions.

The contributions of this group are numerous and critical as they form the key link between the caretakers and the end solutions required within all application domains. The most common and important of these contributions include:

- The translation of domain needs into feature or other technology requirements from future versions of the OS Web mapping technology.
- Influencing the direction of funds to the caretakers for feature advancements, support services and value-added products.
- Training for other stakeholders who contribute to the ecosystem, including the specialists and consumers.

It is tempting to consider professionals as being primarily the domain of consultants, and although many Web mapping consultants do exist, they do not represent the majority of the professionals. Most professionals are employed by organizations that have Web mapping as a key core need for their organization. One of their primary roles is to act as a liaison between the caretakers and the organization they represent.

3.2.4 Specialists

Almost as important as the professionals within the ecosystem are the specialists, who include those who have a rich understanding of a specific vertical market or domain, and who work closely with the professionals to design and deliver high quality Web mapping solutions. The specialists are the stakeholders who are closest to the needs of the end users of Web mapping applications, and in this role, they are instrumental to the successful interactions of the OS Web mapping ecosystem.

Typical contributions of the specialists include:

- Sourcing funds for the activities of the professionals, and providing input to the caretakers for feature advancements, support services and value-added products.
- Expression of real-world needs for Web mapping solutions that are translated by the professionals into technology requirements.

Although the specialists are typically one-step removed from caretakers, it is the rapid and almost seamless flow of information from specialists through professionals to the caretakers that facilitates the advancement of Web mapping technologies that make this ecosystem so effective.

It is here that the difference between a strictly proprietary and OS software development model is the greatest. First, there is a high value placed on the ability to share information quickly from specialist to caretaker and back again. Second, a high value is placed on the service to deliver on these needs in lieu of the actual technology. Together, this seamless interchange from specialist through professional to caretaker is the central nervous system of the OS Web mapping ecosystem.

3.2.5 Consumers

Caretakers, professionals, and specialists each play critical roles as stakeholders in the ecosystem. However, there are many other individuals and organizations that also contribute to the ongoing success of the ecosystem but do not neatly fit into any of these categories.

These stakeholders, generically referred to as consumers, capture the broad portion of the ecosystem that is not always very visible. Consumers are the broad user base of OS Web mapping technologies who do not directly participate in the caretaker roles of software development and support. However, they do contribute in a number of indirect ways. For example:

- Sponsorships, purchase of merchandise, or small donations all contribute to supporting the ecosystem.
- Attendance at conferences and volunteering to organize conferences, without which many conferences simply would not be feasible.
- Promotion of OS Web mapping technologies can lead to more stakeholders of all kinds within the ecosystem.

- For some commercial entities, use of an open source technology means the consumer is not using a competitive (proprietary) alternative.

The sheer number of consumers in the ecosystem can result in major cumulative impacts from what may individually be many small contributions. This cumulative effect helps to shape and direct the ecosystem as it evolves.

3.2.6 The Living and Growing Ecosystem

The OS Web mapping ecosystem is a vibrant, flexible and active environment that is witnessing very rapid changes. The author's own organization, DM Solutions Group, is one of the caretakers of MapServer and MapGuide OS projects and, as such, has had to re-invent itself several times in the nine years of its existence due to the rapid evolution of the OS Web ecosystem structure.

As the tools for collaboration continue to evolve and improve, and the organizations and culture within the community mature, this living ecosystem continues to change at a breathtaking rate and in unexpected ways that are difficult to track even for those at the heart of the system.

From this seeming chaos, three clear trends are emerging:

- Organizations and individuals are defining ever clearer roles for themselves within the ecosystem.
- Adoption is rapidly increasing in a number of communities of practice.
- Businesses models are maturing to capitalize on the opportunities that are created from the multi-dimensional growth of the ecosystem.

The topic of how businesses are succeeding as contributors to the ecosystem is discussed in the following section.

3.3 Business to Business – Business Models Within the Web Mapping Ecosystem

The explosive growth in adoption of OS Web mapping technologies is tightly coupled to a concurrent growth in the volume of business activities surrounding these technologies. For instance, when DM Solutions Group first became involved in the MapServer project as the first commercial caretaker of the technology, total installations were believed to number in the low hundreds. Today, in part as a result of DM Solutions Group's activities, combined with inputs from many other companies, the number of installations is now estimated to be over 100,000 within a seven year time frame.

The primary reason for this growth is that economic incentives drive companies in caretaker roles to complete tasks that individuals within the ecosystem may not be motivated to perform by themselves. For example, one of the first commitments

DM Solutions Group made to the MapServer project was to ensure regular maintenance of a Windows installation of the technology. This was motivated by a need for growth in adoption of the technology that would drive contributions and business opportunities for the company.

The emergence of corporate members of the OS Web mapping ecosystem may not be a surprise. However, the roles that they perform and the dynamics among the corporate participants is in fact very intricate and important to the structure and functioning of the OS Web mapping ecosystem.

3.3.1 Caretakers and Professionals

Caretakers as the primary developers of OS Web mapping technology must, by necessity of their role, focus a great deal of their corporate knowledge, skills, and organizational structures towards their highest value services in support of and advancing these technologies.

Conversely, professionals focus the majority of their equivalent corporate knowledge, skills and organizational structures on services that seek the integration of Web mapping technologies and building solutions for various market segments and geographies. Although caretakers provide significant consulting services to the marketplace in which professionals operate, it is an impossibility for an industry that is broken into many market segments to allow any single company to play a dominant role in all or even a majority of the constituent segments.

As a consequence, what is emerging is a pattern of partnering relationships (both informal and formal) between organizations that are predominantly either caretakers or professionals. This allows caretakers to benefit from business activities in a far greater number of market segments than would otherwise be possible. With technologies that are evolving so quickly it also allows caretakers to stay current with the latest developments and advance Web mapping accordingly. At the same time, professionals with relationships to the caretakers can share in the benefits of these advances.

Although these relationships are often long term, they frequently come and go on a project-by-project basis. This demonstrates the flexibility and resilience of the OS Web mapping ecosystem to embrace change and develop on parallel and often highly integrated trajectories.

3.3.2 Caretakers, Professionals and Specialists

The ecosystem becomes more intricate as well as more valuable when the synergies among caretakers, professionals and specialists combine for shared business gains. Specialists may lack in Web mapping technology expertise, but they more than make up for this in their critical domain expertise for solving specific niche problems. Almost every Web mapping solution requires a degree of technology development or support of some form or another, as well as the expertise to build upon this technology, and the domain knowledge to apply it correctly. When scenarios emerge with a requirement from each of this trio of needs, the ecosystem provides its greatest value.

The nimbleness and flexibility of companies organized by partnerships delivering the caretakers, professionals and specialists to solve a business problem is unique to OS Web mapping technologies. Uniqueness is not derived from the partnership between companies, but from the inclusion in this partnership of a core software development team in the form of caretakers who work in typical software development environments, isolated from the application of technologies to real world problems. In a conventional proprietary software development model, the caretaker company invests the majority of its corporate energy in product development. The consequence of this investment is the loss of opportunities for synergy with other companies. The caretaker focuses on the next release of a pre-defined software package. Responding nimbly and quickly to partner or meet customer needs is secondary to getting the next product release completed and released. However, the requirements for Web mapping are not so easily pre-recognized or defined. They vary extraordinarily among vertical applications. The components that are generic among them all have become commodities to the extent that they have inherently little value.

Caretakers in an OS development model focus on service delivery and they rely heavily on the community for their contributions to common requirements across all users. This difference affords the caretakers the ability to dedicate their best people to their core premium service of advancing technology based on the needs of their best partners and customers. It is for this reason that the power of the business relationships among caretakers, professionals and specialists is unique to the OS Web mapping ecosystem.

In practice, business relationships among the three parties can be very difficult to establish and manage, so more often than not the most successful relationships are between caretakers and specialists who share between them the expertise of the professionals. One of the first and most important roles of the caretakers is to fast-track training and knowledge transfer to the specialists such that they can self-sufficiently perform a professional role within their organization. This role is particularly important around OS technologies as this process enables the specialist company to become far more vendor-independent since more than one caretaker company can and usually does emerge around successful OS technologies.

3.3.3 Servicing Business Consumers

The role of the consumers cannot be forgotten within the context of the ecosystem model. Although consumers may not actively contribute to and engage in the core of the ecosystem, the volume of individuals in this category can create an attractive market segment for the right services and products.

The Web mapping space has not as yet seen a great deal of this development occurring, with a few exceptions such as maps from the MapServer product line from DM Solutions Group and the MapGuide Enterprise and Studio products from Autodesk. However, continued growth, maturing and commercialization of Web mapping will undoubtedly lead to business innovation to fill as yet unknown voids.

The early consumers of standardized products remain at this time "early adopters" of the classic distribution of adopters along the technology adoption curve. This group of consumers includes representatives from government, NGOs and commercial sources. As the availability of standardized products and services become more mature, this will ultimately lead to much broader consumer adoption in the early majority and ultimately late majority stages.

Consumers also play another important indirect role in the process described in this chapter. Specifically, by leveraging OS Web mapping technologies and not purchasing competitive products, the competitive space is altered among software vendors and providers in such a way that can have a dramatic negative impact on proprietary solutions, and provide a positive impact on OS-based solutions.

3.3.4 Partnerships – The Glue of the Ecosystem

Powerful partnerships, though in their infancy in the OS Web mapping ecosystem, are increasingly bringing together the strengths of the key stakeholders to solve business problems. As business relationship innovations transform into repeatable best practices, partnerships may become the single greatest catalyst for propelling OS Web mapping technologies into being the dominant technologies in the spatial information technology industry.

The speed at which OS web mapping has emerged from an almost immeasurable component of the industry to a recognized force today underlines the pervasiveness of its growing cohesiveness. Clearly, the power of business relationships around the related technologies is not speculation or imaginary. It is reality. However, partnerships are not limited strictly to business-to-business relationships. Other relationships between entire communities of practice and industry also define the OS Web mapping ecosystem.

3.4 Business to Communities of Practice Business Models

As previously discussed, the traditional proprietary software model is problematic with respect to Web mapping due to the varying demands placed on Web mapping from so many different industries. The result of this disconnection is that technology is produced that addresses only the basic needs for any one particular market segment rather than having the flexibility to cover multiple market segments at any one time. Businesses that deliver customised solutions to niche needs take a different approach.

The caretakers recognize from the start that OS software will be deficient for any particular market segment, so the first question becomes how can this need be adapted to meet a particular community's needs? The OS approach avoids the product development pitfalls of the proprietary software developers by making the core software freely available to the person, group or organization that wants it, and most importantly to a diverse community to help build it. Suddenly, the relationship between user and supplier changes enormously. No longer is the relationship passive, as it is in the proprietary world, but it is active and dynamic where producers can be consumers at the same time. This creates an environment where license fees or usage restrictions or software shortcomings are not impediments to progress. Instead, collaborators work together to develop needed solutions that then become accessible to other developers and consumers, and so the cycle continues without impediment.

Different motivations drive the two types of software markets (OS and proprietary). An OS supplier measures success not by how many software licenses users buy, but by whether the solution meets their needs. Eliminating license fees for commodity technology from the equation dramatically alters the relationship with vendors, as the focus changes.

This approach to technology development lends itself particularly well to commercial entities developing OS technologies to work with communities of practice (CP). CP are drawn to Web mapping for its potential to solve important problems within their domains, and to deliver information in a manner that is unique from other technical mechanisms. As discussed previously, it is the application of these technologies that crowns their value.

CP also often contain individuals who act in the professional role described previously in concert with the many specialists that exist within the community. This focuses the need for specific capabilities and requirements from the underlying Web mapping technology and forms the foundation for the partnering relationships that often thrive between the caretakers and CP. The existence and nature of these relationships are common requirements to advance OS technology. Where the needs of the community and the needs of the caretakers merge, synergies and partnerships occur. These relationships are often associated with directed funds for accelerating technology development, a proven user requirement, and a community ready to build, test and document the resulting technology. All parties win, the CP, the caretakers and their customer base, and all existing and potential users of the underlying technology.

CP can often be highly distributed spatially and vertically without a strong internal organizational structure. However their collective needs from a technology perspective can be strong and valuable enough to drive important business relationships that seek to advance the use and nature of technology. Without the onus of licensing fees to restrict uptake, resources can be directed at the services to ensure that the technology meets its specific requirements.

3.4.1 Communities as Drivers of Technology Requirements

Sometimes an organization or user community within a CP will enhance OS technology to address a specific need. In that case, typically the developers pay for the enhancement by developing it in-house or contracting it's development to a third party. However, the enhancement then becomes available to everyone else for free if it is built under an OS license. For instance, DM Solutions has done a lot of work enhancing MapServer technology to conform to Open Geospatial Consortium (OGC) standards. Partnering with GeoConnections, an initiative within the Canadian Geospatial Data Infrastructure (CGDI – see Chap. 1), to make Canada's geographic information available on the Internet, DM Solutions improved the MapServer project to support OGC standards and to conform to Canada's interconnected network of spatial databases.

This work attracted the attention of the Southeastern University Research Association (SURA), a consortium of more than 60 universities across the southeastern United States. SURA is developing prototype applications for an integrated oceanographic observation system that will rely on many of the OS technologies and open-standards concepts that the CGDI employs. Web mapping and other GIS technology producers have overlooked the oceanographic community, largely because oceanographers need to analyze wind speeds, current velocities, temperature changes and other phenomena in relation to time, an attribute that hasn't much concerned the GIS world until recently (see the development of TerraLib and associated tools discussed in Chap. 12). Consequently, GIS technology has traditionally failed to address the oceanographic community's needs.

To fix this shortcoming, SURA hired DM Solutions to include a temporal dimension in MapServer software and improve the display and querying of raster imagery to satisfy the basic data requirements of oceanographers. Now the entire oceanographic community, in America as well as worldwide, can use this enhancement to MapServer for free. SURA funded the development of the application, but the enhancement now is part of the OS platform within a specific CP. This dynamic is one of the main attractions of the OS model in that as soon as one community enhances the technology, every other community, whether directly related or not, can use the enhancements at no cost other than the time and expertise required to deploy it and use it productively. The development costs are typically equivalent to the proprietary software market with the key difference being the lack of purchase and licensing costs to the end user.

3.4.2 Business Working to Meet Community Needs

DM Solutions has also harnessed the same technology advancements discussed above to spawn further feature advancements within the MapServer project, to work with partners and customers on novel public and commercial applications, to advance awareness of the company and to help advance the uptake of the technology that forms the core of its business. In this process the core Web-based mapping technologies have diffused widely in Canada as well as in North America in general and around the world. As the momentum of adoption of the technologies continues, so does the drive for collaboration among industry and other CP to meet their own (and others) needs. The cycle continues to build upon itself with new enhancements of the technology, creating more opportunities for communities to meet their goals, and business to support them in this effort.

Using this recursive approach to mutual strengthening and support, CP have become a central building block of the OS Web mapping ecosystem. Representatives of these communities are often key contributors of code, testing and expertise within the OS software community, broadening even further the role and strength of the ecosystem. Communities, though, are not silos and one of the more fascinating consequences of CP being active in the ecosystem is the ways in which they work for mutual benefit with completely different communities living within the same application domain or within unrelated application domains.

3.4.3 Communities Helping Communities

The following scenario indicates how the contribution of one community can benefit another. By enhancing a software project, the appeal of the underlying technology is widened, and this increases the likelihood that another community will support an enhancement that benefits the first community, and so on. For example, Environment Canada needed to publish point data stored in a variety of database formats to OGC specifications. By funding DM Solutions to add the capability to read point data from any database format through MapServer, Environment Canada was able to offer this capability to all of its employees and to any other community with similar requirements.

The client in this case could not have predicted that this addition would compel SURA to adopt MapServer technology, but it did. In turn, SURA funded MapServer's time-feature enhancement, as noted earlier. In turn, Environment Canada has capitalized on the SURA-funded enhancement regarding temporal data to publish time-sensitive data to OGC specifications. In the future, another organization can build on these enhancements, and the process continues to iterate to the benefit of yet unconnected CP who will enter the ecosystem. The OS model invites this continual improvement, which makes the underlying technology more versatile, more timely, more responsive to practical needs and, ultimately, more valuable as a resource.

All of this takes place because a great deal of embedded value exists for applying the technology to solve a business or community problem. The cost of a simple enhancement is insignificant compared to the net impact this has on the resulting application of the technology. For this reason the ecosystem thrives.

3.5 Empowering Communities of Practice and Business Through Open Source Web Mapping Technologies

All of the scenarios, participants and activities described in this chapter are currently working together and growing in momentum as each technical advancement and business innovation broadens the appeal of both OS technology and the businesses that are designing business models around it (see Chap. 2). Together, these forces combine on a formidable progression towards broad adoption of OS Web mapping technologies within the spatial data industry. To summarize these forces, they include:

- *Technology Maturation* – a result of growth in corporate and contributor experiences combined with business dependence on the technology
- *Business Model Maturation and Innovation* – all drive new and more commercial relationships to OS technology
- *Growing Business Relationships* – Partnerships take time, and with each passing year more emerge and strengthen
- *Broadening of Product and Service Offerings* – maturing business models result in greater selection and choice of means to interact with OS technology
- *Capacity building and Maturing of Communities of Practice* – greater capacity within communities of practice result in further opportunities to advance OS technology and its adoption.

3.5.1 A Big World, Varying Geographic Requirements and Common Geographic Capabilities

In addition to the many relationships discussed in this chapter, and the specific forces of momentum, one other unique aspect of Web mapping lends itself towards supporting the evolving ecosystem. This aspect is the very nature of geography itself.

The world we live in has immense diversity and in no way does this manifest itself more clearly than in the manner in which places are described. Whether information or patterns are shaped in scientific, cultural, socio-economic or other terms they all share a common identity in geography. Just as Web mapping, when used in various domains, requires the application of technology to capture value, geographic variety also drives value into the use of information that is independent of the technology itself.

This worldwide variation in how value is captured in the application of information with OS technology provides an added accelerant to the forces supporting the growth of the OS Web mapping ecosystem. This is primarily revealed in the roles of the professionals who must be well versed in capturing and understanding geographic variability in addition to understanding the nature of application needs. This, again, impacts on the role of the caretakers, who must accommodate within the technology the specific needs of varying geographies through relationships with professionals. These dynamics would be otherwise very difficult to achieve within a traditional proprietary software model given the closed nature of the relationship cycles.

Finally, variations in geography lend themselves to needs being met locally rather than internationally, especially in the context of professionals and specialists. Nonetheless, the role of caretakers remains critical in order to deliver high quality solutions. Hence, the result is strong partnerships formed due to geography. As has been seen in this chapter, the unique nature of partnerships between caretakers, professionals and specialists, and how these fit the OS Web mapping ecosystem so powerfully, naturally extend themselves into geographic relationships.

3.5.2 Building Communities from Mutual Successes

Ultimately, a strong ecosystem, which fundamentally is its own powerful community, only finds strength because it's members achieve greater success by being contributing members than by acting in isolation. In this respect, the nature of the OS Web mapping ecosystem is very much a symbiotic one. Whether for personal satisfaction, for risk management and aversion within a CP, or the advancement of commercial goals, all members of the community share the same goals, without which the ecosystem would never be sustainable.

Today, many of the challenges within the ecosystem are from the necessary turbulence that must take place as the community matures, as new individuals and groups begin to participate and inevitably some of the original participants may no longer have a role. It is all part of a natural evolution that must take place for the overall health of the ecosystem and its sustainability to thrive.

The shared drive for success among all contributors and the often highly interlinked relationships that demand success from each other create a sense of partnership and joint ownership that is unique to OS technology.

3.5.3 Nurturing and Growing the Open Source Web Mapping Ecosystem

In most respects, the ability to grow and strengthen the ecosystem lies in the simple actions of each individual and organization collectively acting in self-interest to maximize their success with OS technology. This has driven most of the contributions and activities that have achieved the successes of the ecosystem to date. There have been a few notable exceptions to this rule and undoubtedly there will be more in the future.

Some activities are a step removed from directly benefiting a specific organization, but are fundamental to the long term health and success of the ecosystem. One of these moments arrived with the formation in 2006 of the Open Source Geospatial Foundation (OSGeo).

In order for OS geospatial technologies to continue to progress in their adoption and credibility within the geospatial industry an arm's length entity was required to ensure best practices were observed for the development of these OS technologies, and to form the rules by which individuals and organizations could participate and influence the future direction of technology. Although not directly of interest to any one party, the collective benefits of making this happen drove contributors from throughout the FOSS4G technology space to participate in the formation of the foundation. None of this would have been possible if the ecosystem and broader participants did not have an existing community of trust to work from.

Another example of this is the formation within many OS technology projects of steering committees for decision making in terms of technology direction. This important step, again not set by any one particular participant but on the basis of the collective good, is achieved in many of the key ecosystem projects due to the trust and relationships that exist within the community. These steps and many others will be required in the future as the ecosystem continues to mature.

Clearly, the most important activity that all participants within the ecosystem can and must strive for in order to achieve mutual success is growth in trust and development of a sense of community within the ecosystem. It is this thread that will be at the heart of the ecosystem of the future, whatever it may look like. Given the strength this community has shown to date, and the speed at which it has formed and formalized, no-one should doubt that it will succeed.

Chapter 4
MapServer

Stephen Lime

Abstract MapServer has emerged from National Aeronautics and Space Administration (NASA) sponsored research at the University of Minnesota to be become one of the most popular Open Source GIS packages available today. The software provides users with the basic tools necessary to build spatially-enabled Web applications using their own data. MapServer integrates with other popular Open Source applications such as PostGIS, Ka-Map and OpenLayers and supports most major Open Geospatial Consortium (OGC) Web service specifications. This chapter details how MapServer came to be, it's strengths and weaknesses and where it is headed in the future.

4.1 Introduction

The University of Minnesota MapServer project (henceforth referred to as "MapServer") has been a fixture in the free and open source for geospatial (FOSS4G) space for a decade. MapServer is a platform for developing spatially-enabled, Web-based applications and services. The software, written in ANSI C, runs on just about any computing operating system, from AIX to Microsoft Windows.

MapServer comes with an out-of-the-box common gateway interface (CGI) application that provides functionality to build interactive Websites, Web 2.0 application components, and Open Geospatial Consortium (OGC) service instances. It also features MapScript, a powerful scripting interface for popular languages such as PHP, C#, Java, Perl, Python and Ruby. MapScript allows developers to add geospatial functions to any application.

Stephen Lime
Minnesota Department of Natural Resources, St. Paul, MN, USA,
e-mail: steve.lime@dnr.state.mn.us

G.B. Hall, M.G. Leahy (eds.), *Open Source Approaches in Spatial Data Handling*.

4.2 Project History

The MapServer project certainly did not begin with any notion of Open Source (OS) software in mind. In fact, the concept of or term OS was not even in widespread use at the time of MapServer's initial development. The project was born of necessity and frustration with commercial Web-based mapping offerings available in the mid-1990s. The initial interest was in developing spatially-enabled Web applications and not the software to make that possible. The following sections detail the events that lead to the creation of MapServer.

4.2.1 Boundary Waters Canoe Wilderness Area

The ideas that formed the basis for MapServer were created as part of an effort to develop a decision support system for recreational users of the Boundary Waters Canoe Area Wilderness (BWCAW) in northern Minnesota. The BWCAW is a collection of lakes and campsites scattered over 1 million acres and connected by a network of streams and portages. The "routes" that visitors might take on a trip could vary in a number of ways, such as degree of difficulty, species of fish to catch, or opportunity for solitude. The initial idea was simply to use the Web and geographic information systems (GIS) technology to fashion a tool to allow visitors to identify routes based on their needs. In this context, the Web provided generalized access for multiple users without the need for them to own or have knowledge of GIS software in order to establish canoe routes for wilderness interaction.

The initial implementation used a basic configuration file in combination with user input via a Web browser to author Arc/Info Arc Macro Language (AML) code. This code was subsequently submitted to workstation Arc/Info's ArcPlot subsystem for processing and rendering. The resulting graphics file was converted to Postscript using Arc/Info tools and finally to a Web-compatible GIF image, using the freely available Ghostscript software (perhaps this was the OS seed for the eventual growth of the MapServer project into its current state). Feature queries were handled in a similar manner with custom AML scripts created to extract feature attributes for display in the rendered map image. Output was run through a simple, yet highly efficient templating system that is still used by MapServer in its current version some fifteen years later.

This process worked well relative to the initial and highly focused objectives of the initial project. In 1994 users could interact with GIS data through the Web for the BWCAW area and extract simple feature-level data. However, there were two main problems. First, the process was slow, requiring some 30 s to render a map on Sun Sparc2 class hardware. The output was aesthetically pleasing thanks to the conversion through Postscript, but performance response time was clearly not acceptable. Second, there was a licensing issue with the use of proprietary software produced by the Environmental Systems Research Institute (ESRI).

These constraints served as much as anything as the impetus for moving to a more liberated environment consistent with the objectives of OS software in general and FOSS4G in particular. The initial development team had only half a dozen Arc/Info licenses at its disposal at the time of development and losing software access time to the meet the needs of the general public did not sit well with University-based graduate students who had to share software access with wilderness canoeing enthusiasts. In short, there had to be a better way to meet the needs of both groups as well as make spatial data available to multiple users via the rapidly growing Web.

Alternatives, both free and commercial, were immediately sought. At the same time as these initial developments, the Xerox PARC MapViewer was the preeminent Web mapping site but the creator of that service had no interest in distributing the code. There were no other usable commercial solutions at the time. It wasn't until the release of Shapelib by Frank Warmerdam in 1995/96 (see Chap. 5 of this text) that an alternative to the hacking of ESRI's workstation Arc/Info approach emerged.

4.2.2 NASA/ForNet

In 1994, the University of Minnesota, Department of Forest Resources (UMN) won an award as part of the National Agency for Space Administration (NASA) Cooperative Agreement Notice "*Public Use of Earth and Space Science Data Over the Internet*". This project was called ForNet and was a partnership between the UMN and natural resource management agencies, principally the Minnesota Department of Natural Resources (DNR). The idea for ForNet was born from Web-based distribution of forest inventory data being done at the University. In a nutshell, the basic idea was to deliver data, both tabular and spatial, to forest managers in support of on-the-ground management activities. In hindsight, the ideas were somewhat ahead of their time, but regardless the project proved to be fertile ground for research and development of Web mapping technologies.

It was under ForNet that MapServer first took shape in the form that is a precursor to the software of today. ForNet developed two server-side Web mapping products. Given the connection with NASA there was an initial focus on raster imagery and derived products (e.g. classified images), so the first application, called "imgserv", was written using the ERDAS Imagine C application programming interfaces (API). It was implemented as a CGI process and took map extents and band combinations as input to produce jpeg images from Landsat Thematic Mapper scenes. While this approach worked well, it was clear that the imagery by itself was difficult to use, especially in parts of Minnesota that were devoid of landmarks. Consequently, the "mapserv" application was written essentially to add context to satellite imagery. Integration was accomplished on the client-side using Java applets to composite 24-bit satellite images with 8-bit GIS overlays.

In all, a dozen applications were built as part of ForNet including:

- Satellite image catalog and image browser
- Wildfire location and status mapping
- Multi-image channel change detection browser
- Forest stand inventory access system
- Digital aerial photograph distribution system.

Several of these, albeit modernized, are still in production use at the DNR today.

4.2.3 State of Minnesota, Department of Natural Resources

At the time the ForNet project concluded MapServer use was limited primarily to UMN applications on a handful of projects. There were a few external organizations using the software (primarily overseas), but the user community numbered less than 10. Version 2.0 was released as a final project deliverable.

In the fall of 1997 MapServer development moved largely to the State of Minnesota DNR. The DNR took ownership of most of the applications developed under the ForNet project and as such had a vested interest in continuing their development.

Several key features were added to the software during this time period that represented significant improvements over the Version 2.0 feature set. For example, basic raster support was added so it was possible to integrate raster and vector data sources on the server side. While this approach worked well (it allowed 24-bit raster presentation with an 8-bit overlay), it was becoming increasingly difficult to rely on applets because of issues with Microsoft's support for Java. Another major addition was support for Truetype fonts, first natively via Freetype (http://www.freetype.org/) and then directly through the dynamic graphics generation GD library (http://www.libgd.org/). Label collision resolution was supported through a label caching mechanism that proved very effective in placing as many labels as possible (and quickly) without any overlap. Automatic label rotation meant that road networks could be labeled without pre-processing so large national datasets like Tiger or that produced by Geographic Data Technology (GDT) could be rendered in MapServer to produce street maps suitable for locating spatial features.

4.2.4 NASA/TerraSIP – MapServer Goes Open Source

On the heels of the ForNet project, the University of Minnesota was once more funded by NASA to deliver land use and land cover data over the Internet to land managers. This project, called TerraSIP, began in early 1998 and used MapServer as a tool for information dissemination. Project research focused on using multi- and hyper-spectral data within the MapServer framework. Ultimately that research didn't get added to the MapServer code base (given the availability of GDAL – see Chap. 5) but the work certainly influenced future investment in development of MapServer. In early 1999 the University was approached again to play a central role with MapServer software. The decision to open the source code for MapServer was made at this time and project infrastructure was deployed to this end. TerriSIP supported the code cleanup and documentation for the first Open Source (OS) release of MapServer 3.0 later that year.

Several point releases occurred before MapServer Version 3.3 was released with an important new capability called MapScript (described in detail below). This version caught the eye of Daniel Morissette of DM Solutions Group (see Chap. 3) who made initial contact to investigate whether additional contributors to the MapServer OS project would be sought. The response was affirmative and DM Solutions offered two significant additions: (1) to develop a PHP-based version of MapScript and (2) to make MapServer run under Microsoft Windows. These were key developments in terms of broadening the appeal of MapServer beyond the Unix/Linux market where it was popular at the time.

4.2.5 *Why the MIT/X11License?*

People tend to be very passionate when it comes to the licensing of OS projects. The approach used for MapServer licensing was simple, namely for a new project to be successful there should be as few impediments to adoption as possible. The MIT/X11 license (http://en.wikipedia.org/wiki/MIT_License) says what it needs to with regard to use by third parties and it allows developers to build upon the foundation architecture by adding new functionality and enhancing the capacities as well as robustness of the software. It was felt that if MapServer was of sufficient quality then users would want to give back to the project by providing enhancements, since users are those who tend to be best suited to know what they want with a tool that they use. The freedom the MIT/X11 license provides in this regard has increased use of the software and propelled it into realms that were not initially envisioned in the application developed for the BWCAW area.

4.3 Design/Development Philosophy

MapServer is different from many projects because it was not the product of a formalized development process. On the contrary, MapServer is the product of countless hours of prototyping, of writing and re-writing, and trial and error, especially early on. Instead, functionality evolved organically. When something new was needed it was simply added, either via new code or by integrating an external library or two. This is certainly not the most efficient means of developing software, and in fact a great deal of energy has been spent in overcoming some of the early design decisions (see Sect. 4.6).

4.3.1 Leveraging Best-of-Breed Open Source Tools

Conceptually MapServer is actually very basic. However, Web mapping and GIS are not simple, either conceptually or operationally. The act of rendering a map on the Web requires a large number of specialized functions including 2-D graphics, GIS file I/O, cartographic projection, font rendering and many others components.

Clearly, it is not practical for a single project to develop all of these capabilities by itself. Fortunately, the OS world is replete with niche libraries so that developing MapServer really became a discovery and integration task as much as anything else. Important questions such as "how can we make geometries that are read from ESRI shapefiles using Shapelib, transformed using Proj. 4, and ultimately rendered using GD?" need to be addressed. To achieve these sorts of tasks, MapServer relies on more than a dozen supporting low-level libraries and development tools. Without these there would be no MapServer. Hence, this is the essence of OS development.

4.3.2 "No Code" Web Development

From the perspective of a C programmer it was apparent early on that for MapServer to be successful it needed to provide a development experience that required no programming. As a result, MapServer relies heavily on the use of external configuration files (a mapfile) and presentation templates containing special tags that are intuitive and easy to use. For example, the out-of-the-box CGI program supplied with MapServer provides a significant amount of functionality and minimizes the need to write server-side code. Hence, it is still possible to develop a typical pan/zoom/query application with only a mapfile and an HTML template or two.

4.3.3 Think-UNIX Pipes

Building on the no code theme, MapServer drew upon the UNIX operating system and its "pipe" operator for inspiration. MapServer provides the ability to be able to chain requests to the CGI together to create sophisticated output. For example, an initial request could start with a mouse-click in a map of polygons. The geometry of the selected polygon is then embedded in a subsequent call to find all points in another layer contained in the polygon. The resulting set of points is displayed with inset maps (generated with a call to the CGI) showing the selection polygon and the selected point. No programming is necessary, just an understanding of how templates and calls to the CGI interact. Once mastered, this understanding can open many other possibilities for developing practical and easy to use applications.

It is possible that many Web-based applications can work in a similar fashion, but MapServer was designed explicitly with input and output capabilities that facilitate the linking of small, simple pieces of functionality into larger end-user driven applications.

4.3.4 Simple Features and Functions

The earliest functional requirements for MapServer were quite meager. The software needed to be able to produce an image graphic, as a map, based on some description of an area of interest, typically given as a bounding box. There needed to be a means of interacting with the map to navigate, pan and zoom. Finally, a user would need to be able to identify a feature using a coordinate and retrieve some information about it. Hence, three basic functions, namely pan, zoom and identify, form the core functions of the software.

These functions were initially encapsulated in a single CGI program mentioned previously (see Fig. 4.1). By changing a single parameter called "mode" it is possible to turn a user's mouse click into to a new map or use that click to select features from one or more spatial data layers. In fact, it is amazing how many problems can be addressed by these core functions.

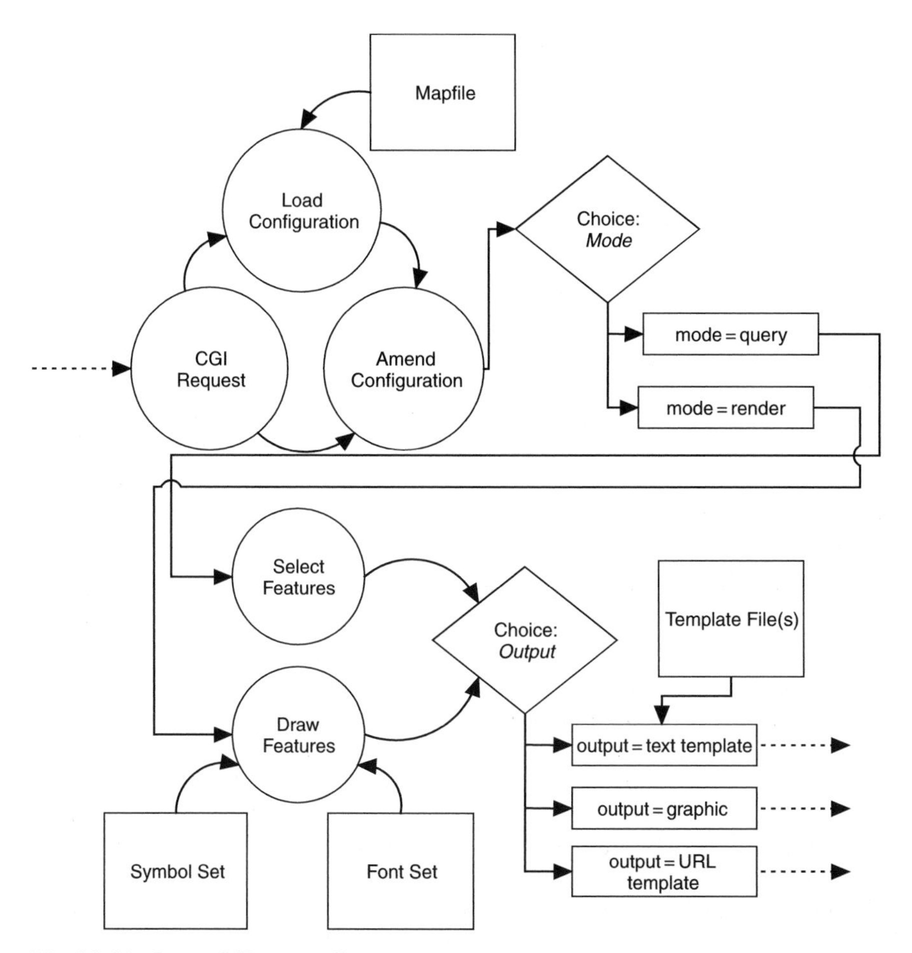

Fig. 4.1 MapServer CGI process diagram

4.3.5 Making Maps

The implementation of the above functions sounds simple enough at a first glance. However, making a map is not all that straightforward in practice. There are a host of complications, including reading spatial data formats, coordinate transformation and projection, configuration management, symbol storage and rendering, and feature annotation to name a few. Details follow on how MapServer manages these tasks. From a design perspective, the goal of the software has always been to produce beautiful maps as fast as possible.

MapServer has the ability to create maps as part of a larger application so that images are written to a writable directory and referenced in templates, or it can act as a map engine and return an appropriate mime-type and binary stream directly to the client browser. The latter is a useful feature for embedding links to dynamically created maps using a simple HTML image tag.

4.3.6 Map Element Automation

As soon as a map is created it is natural to want to know what it represents, the size of the objects that are represented in it, and where it sits relative to the larger picture (e.g. a state, a county, or the world). As a result, MapServer supports the automatic generation of components a user would normally find on a map including legends, scale bars and reference or key maps. In addition, these elements (primarily scalebars and reference maps) can exist independently from one another so that, for example, MapServer can be used only as a scale bar engine.

4.3.7 Feature Query

While at its core MapServer makes maps, it also contains a reasonably powerful feature query capability. Originally, MapServer only required point selection functions, that is, click in a layer and return the attributes of the selected feature. Eventually, however, other project work (for example, selecting Landsat images from overlapping coverage boundaries) required being able to identify multiple features in one or more spatially coincident layers from a single point. It turned out that the process of identifying features for rendering is essentially a bounding box query so that was added next. Several additional variations have been added over time, including query by attribute and by arbitrary polygon as primary examples of these.

4.3.8 Presentation Layer Templates

When using dynamic spatial data, map elements and query results need a method to make this information available to the user. As mentioned previously, the MapServer CGI uses a templating system to accomplish this.

Depending on the way the software is called, a variety of tags are available to the application developer to expose any number of details about the processing that has just occurred. For example, in a pan/zoom request the tag [img] is used to retrieve the name of the new map, [scalebar] is used for the scale bar image, and so on. Information about the extent and size of the map are also available. MapServer supports a limited number of tags designed to help maintain the application state, that is, what layers were requested and was the user zooming or panning at the time of the request.

When returning a set of query results, MapServer uses one or more templates to present what sometimes can be a very complex and diverse set of information. A result set can contain one or more features for one or more layers. A series of header and footer templates can be (optionally) used to frame individual feature information. For example, it would make sense to output a layer title before listing each line of detail. At the feature detail level, any feature attribute may be retrieved using a special tag that also supports several encoding methods depending on how the data are to be used. Spatial characteristics of a feature bounding box, centroid and even its geometry can all be accessed using templating.

While HTML is used most frequently as a template format it is possible to create other types of output such as plain XML, KML, SVG, GeoRSS or JSON. MapServer also supports a special "URL" template that can be used in situations when only a single query result is to be generated. In this case, tags in a URL are expanded based upon the attributes of a feature, and the requesting agent is redirected to the completed URL. A simple example would be a map showing Minnesota State Parks and when a user clicks on a park they are sent to an existing Web page with more information. Hence, this approach represents a sort of geographic image map.

4.3.9 OGC WxS Specification Support

The Open Geospatial Consortium (OGC) Web service specifications (WxS) provided MapServer an excellent opportunity to integrate capabilities far beyond the powerful, yet esoteric CGI approach. As luck would have it, the major service specifications map directly to major MapServer functional areas. These specifications include the Web Map Service (WMS) for map creation and simple query, Web Feature Service (WFS) for feature query, and Web Coverage Service (WCS) for map creation, with GDAL used as the processing back-end. In all cases, the MapServer CGI backs the OGC service implementations. As a result, when built with appropriate services enabled, the MapServer CGI application will respond to normal CGI requests and/or OGC service requests (see Fig. 4.2).

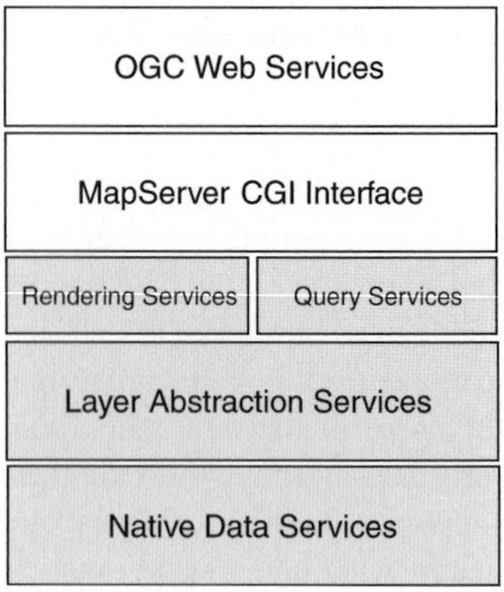

Fig. 4.2 MapServer service stack

Probably the largest gap in functionality between MapServer and any of the service specifications was the fact that OGC compliant servers must be self-describing (e.g. REQUEST=GetCapabilities). MapServer configuration files (described later) initially had no means to store layer metadata (e.g. title, keywords, abstract). Hence, metadata blocks were added to support ad hoc data storage at several scopes including service and layer. Metadata are used to control everything from basic data metadata to service metadata (e.g. supported projections or output formats) to output configurations (e.g. data transformations such as aliases, grouping, and constants necessary to support geographic markup language (GML)).

With a general mechanism to configure OGC services defined, a number of specifications were implemented over the following years. In Version 3.5, WMS support (for both client and server) came first since it most closely matched core MapServer functions. Versions 1.1.1, 1.2 and most recently 1.3 are supported. The useful, but little used, Web Mapping Context (WMC) was supported beginning with MapServer 3.7.

WFS 1.0.0 support (also for both client and server) was included with the release of MapServer 4.0. WFS support is easily the most difficult from an implementation standpoint and so it continues to be the specification that undergoes the most frequent update. Early versions of the WFS code produced only vanilla versions of GML 2.0 documents. Subsequent support has resulted in features such as GML 3.x support, on-the-fly transformations for the production of GML and accompanying XML schema, referencing external application schema, and full topology operator support for filter encoding (e.g. WITHIN, INSIDE).

WCS 1.0.0 support (server only) was the last major specification to be supported with MapServer Version 4.4. WCS is basically a wrapper for the GDAL utilities (see Chap. 5). As a result, anything that GDAL can access can be delivered via WCS. The most recent addition to MapServer is for the Sensor Observation Service (SOS) specification, which was added at Version 4.10.

The OGC support within MapServer is by no means complete. In fact, only WMS 1.1.1 support has passed the OGC certification test suite. Certification is a time consuming and expensive process and hasn't yet been pursued. MapServer OGC implementations can be considered complete, however they are occasionally quirky, with limited support for the more esoteric features.

4.3.10 Spatial Data Format Support – Leveraging Middleware

As mentioned earlier, the event that really started MapServer as it exists today was the release of Shapelib in the mid-1990s. Since then MapServer has always supported ESRI Shapefiles natively. Vector data access was abstracted late in Version 3 with native support for ESRI SDE, PostGIS and Oracle Spatial was added soon after. On the positive side, native drivers performed exceptionally well and could be fine tuned for MapServer use. However, invariably these become difficult to maintain. In addition, the question of where does the development process stop has been ever present in the ongoing growth of the software. Writing native drivers for the myriad of possibilities could become a project of its own. Luckily, at the same time the drivers for spatial databases were written, a driver for the middleware library OGR, developed by Frank Warmerdam, was also added. OGR allows MapServer access to any data format that OGR can read, including 29 formats and counting at the time this chapter was written. In cases where the translation overhead of OGR is unacceptable MapServer supports a plug-in architecture for integrating external third-party data providers at runtime.

Similarly, the first OS releases of MapServer had basic raster support by means of native drivers for TIFF, GeoTIFF and a few GD graphics formats. The raster translation library GDAL was added to MapServer at Version 4.0 and allows access to around 60 different formats. Raster translation does not suffer from the performance constraints that vector translation does, so the GDAL raster driver is the primary driver that is used by MapServer. One important point to note here is that when GDAL support was added, it was also applied to the production of maps and MapServer could write any format GDAL supported (keeping in mind that MapServer rendered 8, 24 or 32-bit images).

4.4 MapServer Architecture and Object Model

The aversion to having to write code to develop early MapServer applications necessitated the need to support relatively complicated external system configurations. Text files were ideal in that they are simple to create, are readable by humans and relatively easy to parse using C (using Lex/Flex). With a good text editor it is possibly quickly to get underway.

4.4.1 Mapfiles

As noted earlier, configuration of a MapServer application consists of a single text file called a *mapfile*. A mapfile is hierarchical format consisting of a series of object definitions that describe an application or service, the data to be used, presentation definitions for the data, and finally additional elements that may be part of a service or application (e.g. scalebars, legends and reference maps). The structure of a mapfile is not unlike XML. It consists of tags or identifiers to start object definitions and the keyword END to close an object. Mapfiles are case insensitive and are relatively freeform within an object definition. Linefeeds have no meaning so it is perfectly legal to define entire objects on a single line. Comments are denoted with a # and quotation marks are used to delineate data containing special characters. For example:

```
# Simple Mapfile Definition
MAP
 NAME 'myMap'
 EXTENT 50000 100000 60000 101000
 SIZE 500 500
 UNITS METERS
 LAYER
  NAME 'myLayer'
  DATA 'myLayerData' # no .shp extension
  TYPE POLYGON
  STATUS ON
  CLASS
  STYLE
     COLOR 255 0 0
  END
 END
END
END
```

The example mapfile shown above defines a 500×500 pixel map containing a single polygon layer drawn with a solid red fill. Adding additional data to the map is a matter of adding additional layer definitions to the mapfile. The order of certain types of objects is significant. For example, layers are drawn in the order that they appear in the file. So raster layers are typically added first, polygon data next, and finally line work, points, and annotation.

The structure of a mapfile closely matches the internal C structures. Figure 4.3 shows a high-level object diagram for MapServer configuration. It looks more complicated than it really is conceptually because so many objects share fundamental things such as colours or labels. In reality it is relatively simple, using the map object as the main container storing a few properties of its own, while most of the detail is held in sub-objects. The following section discusses these sub-objects in more detail.

4.4.2 *Layers, Classes, Labels and Styles*

The concept of a "layer" is central to most GIS, and MapServer is no different. A layer holds all the information necessary for the software to access, query, and

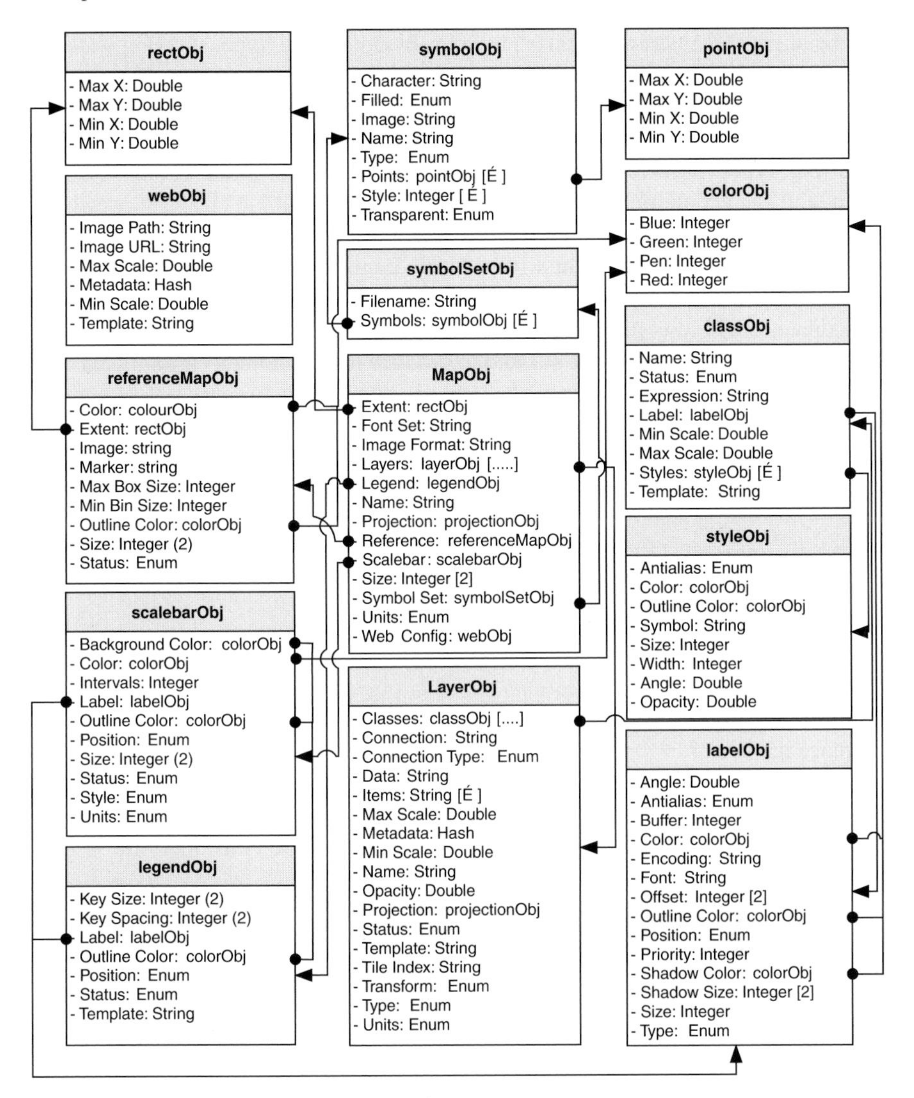

Fig. 4.3 MapServer object model

render features. In most cases a layer refers to a GIS dataset that will be drawn in an image buffer. Data can come from external file-based sources (e.g. Shapefiles or TIFF images), a spatially enabled database (e.g. PostGIS and PostgreSQL) or features defined within the mapfile itself.

To determine which presentation definition (e.g. colour, symbol, size) is required a “class” object is used. All layers being drawn must have at least one class defined. Classes store map presentation definitions in style and label objects and query output definitions in template references. Most importantly, classes are home to expressions, which are used to group features logically for drawing. For example, it may be of interest to draw land use polygons based on their type, showing water in blue, forest in green and so on. Class expressions allow for just that. Three types of expressions are supported, namely straight string matches on a single feature attribute, regular expressions matches on a single feature attribute, and logical expressions (e.g. type="hardwood forest" or type="conifer forest") that use a SQL-like syntax and can act on multiple feature attributes.

Style and label objects contain the actual definitions used by MapServer rendering functions to colour pixels in a map. Styles control which symbol, what colour to use, and how big to make it, while labels control how any annotation associated with a feature will look including font, size, colour, etc. A class can contain several styles, so in effect styles can be stacked to produce more complex cartographic output. For example, the hollow line effect seen in many contemporary online maps is produced by laying down a thick line of one colour followed by a second thinner line drawn in another colour. The order of styles is therefore important, as the first style defined is rendered first and so on.

4.4.3 Map Elements

The map object also holds definitions for other objects that might normally be seen as part of a map or a mapping application. These consist of scalebars, legends and reference maps.

A scalebar definition includes attributes like the base size of the scalebar, any colours or fonts to be used, the number of intervals and output units. Scalebars can be created as stand-alone graphics or they can be embedded into a map on-the-fly.

MapServer legends take much of their configuration from the layers, classes and styles described above so a legend definition is fairly minimal. It consists primarily of parameters to control the size and spacing of key boxes and any fonts that might be used. MapServer supports two types of legends, namely an image-only version that, like scalebars, can be stand-alone or embedded and a text legend. Text legends most often take the form of HTML and are produced using a legend template. In this template a developer can access components such as key images, layer and class names, scale hints and all layer metadata. HTML legends are particularly useful for building custom application layer controls (e.g. collapsible dynamic HTML (DHTML)).

Reference maps are small "locator" maps that are intended to help provide a frame of reference when a user is zoomed into an area. These are created from a predefined base graphic with a known size and extent (often produced by MapServer). Rectangles or markers representing the location of another map are then rendered on top of the base map.

In all cases, map elements can be generated individually (e.g. mode=scalebar) using the CGI interface or as a graphics group in a full pan/zoom application.

4.4.4 Symbol Sets

One of the biggest challenges in defining the configuration for MapServer was for symbology. At the time symbols were first developed there were no de facto standard for representing, for example, a five-pointed star or a dashed line. Only now with the emergence of SVG/CSS and the OGC styled layer descriptor (SLD) specification are there public standards that provide some direction and substance in this regard. The absence of these earlier in the system development cycle represents one of the perils of being an innovator rather than an adopter.

MapServer's symbol format is derived from exposure to capabilities of commercial GIS software but moreso by the capabilities (and limitations) of the GD graphics library. Symbols can be defined in several places including in a symbol set, which is an external file containing any number of symbol objects, as inline symbols, that is, symbol objects defined directly within the mapfile, or implicitly by referencing image files (e.g. GIF or PNG) through the symbol parameter of a style object. In the latter case, MapServer builds symbol definitions on-the-fly as new images are encountered while parsing the mapfile.

MapServer supports 5 basic types of symbols:

(1) Pixmap symbols: images used as markers, brushes or fill patterns.
(2) Vector symbols: inline coordinate sequences used to render markers, brushes or fill patterns. They have the advantage of being scalable.
(3) Ellipse symbols: ellipses (most often just a circle) rendered into a marker or a brush. Most often used to draw wide lines.
(4) Truetype symbols: similar to vector symbols except that the coordinate information is held in a font glyph.
(5) Hatch symbols: a special polygon fill that supports arbitrary hatch angles.

There are a number of secondary parameters such as anti-aliasing, fills, dash patterns that can be applied as necessary.

Different symbol types can be combined through style objects to create complex effects. For example, a Truetype star can be drawn over the top of a circle and used as a polygon fill. For performance benefits MapServer supports a symbol caching mechanism. In most cases a symbol takes the form of an image. Either it starts that way as in the case of a pixmap, or is rendered to one depending on size, colour and angle. The images are stored in a cache, which is indexed by the symbol characteristics so that a symbol at a given size need be only created once per map.

4.5 MapScript

As useful as the MapServer CGI is a developer really needs to think like the author of the software to make use of its full potential. Unfortunately that is not always easy to do. MapScript not only provides functionality beyond what the CGI can achieve. It also allows users to build applications in a way that makes since to them.

4.5.1 MapServer CGI Constraints

As noble a goal as not having to write any code to build a Web GIS application might be, it was clear that the MapServer CGI approach could not be everything to everyone. More and more use cases appeared that identified the need for a tool that allowed developers to access MapServer core functionality under their own terms.

To satisfy this need there is a MapServer C API but its use is not for the faint of heart. It is not particularly well documented, it is prone to change, and it is not terribly consistent. For example, in some cases functions return pointers to locally allocated structures, while in other cases pointers are passed to a pre-allocated structure as an argument to a function that returns a Boolean success or failure code. While efforts are underway to resolve such differences it makes sense to insulate developers from the frailties of this type of environment.

In addition, there are far more potential users for MapServer with experience with high level languages than there are for low level languages, such as C. MapScript provides the mechanism for this high level access.

4.5.2 Scripting Languages

Scripting languages like Perl, PHP, Python and Ruby seemingly provide a perfect means of exposing MapServer core functionality to a large community of application developers. As early as 1997 a trivial PHP module was written that exposed the basic MapServer drawing routine and a few others were created to manipulate map extents and layer state. With this exercise in hand, coding a scripting interface was enough to stop further development of the idea for MapScript. It was a tedious process, prone to error and supporting a single language would be time consuming, not to mention the four or five languages that made sense. At this time, the OS Simplified Wrapper and Interface Generator (SWIG) tool was discovered. SWIG takes C or C++ libraries and generates wrappers in the scripting language of choice for a developer. A relatively simple interface file is written that controls various aspects of the wrapping and SWIG generates the rest.

MapServer Version 3.3 was released with the first version of MapScript with configuration support for Perl (a personal favorite of the author's). Subsequent versions featured bindings for a number of other languages. Currently, Perl, Python, Ruby, Java and C# are all well supported and most other languages SWIG supports have a reasonable chance of working with little or no modification.

PHP presents a special case. SWIG does support PHP very well. So the PHP MapScript binding has been maintained by hand since the mid-3.x versions of MapServer. Hence, PHP/MapScript represents the by far the largest user base of the MapScript languages.

MapScript is important not only from a functional standpoint. It has had a huge effect on the internal operations of MapServer as a whole. As a purely CGI-based package, the underlying C API was free to contain a certain amount of sideways movement. Awkward interfaces are not an issue because no one would see them. However, the arrival of MapScript changed all of that. Lots of wrapper code could be written to mask these CGI-based shortcomings but that approach defeats the purpose of using a tool like SWIG. It is much more desirable to fix the underlying C code so that MapScript falls naturally from it.

4.5.3 MapScript Functionality

The easiest way to understand MapScript is to realize that with it a developer can re-write the out-of-the-box CGI interface with it. The CGI is written on top of the C API, as is MapScript. For example, reconsider Fig. 4.1. MapScript can access functions in any of the boxes (with the exception of the CGI interface). All of the relevant drawing, query, data access and presentation methods are available to the developer.

In addition, there are certain functional areas that are available in MapScript and not through the CGI application. The largest of these is the exposure of GEOS analytical functions (e.g. buffer, centroid, convex hull) and geometric operators (intersect, union, difference). When compiled with the GEOS library, MapScript possesses GIS-level functionality.

The true power of scripting languages is that developers have so many other toolsets at their fingertips. There are modules for everything so it becomes trivial to, for example, retrieve a Web page, parse a coordinate from the content, buffer the coordinates, query a data layer with the resulting polygons, and map the results. There really is no limit to what is possible.

4.6 Future Challenges

The biggest challenge with a project like MapServer is setting a long-term direction and sticking with it. Given the variety of capabilities available in other solutions it is tempting to say, "if package X does that, so should MapServer". However, it is important to realize that MapServer is part of the larger FOSS4G stack, of which each component has strengths and weaknesses. With MapServer, the developers must strive to enhance its strengths including rendering, flexible query, OGC service support, and shore up its weaknesses.

Arguably the greatest strength of MapServer is speed. With a properly configured data store it is remarkably quick. In one respect this may be a result of when the software was originally developed as much as how it was developed. On mid-1990 hardware, speed was of the essence. Hence, MapServer featured support for things like tiled data sets (both raster and vector), spatial indexing (including tiles and indexes) and other optimizations. Of course, it doesn't hurt that MapServer is written in C and doesn't rely on layers of abstractions that can otherwise slow things down.

While the overall goals of MapServer remain essentially constant it is clear that the role of MapServer in the marketplace is changing. The following sections discuss areas of particular interest.

4.6.1 Role as an AJAX Service Provider

Early-on in the development of MapServer, a decision was made not to develop the software into a monolithic one-stop-shop application for Web mapping. MapServer does possess the capabilities to build relatively simple user interfaces using templates. However, as time progressed, the need for more sophisticated application interfaces increased and a number of excellent client-side application development systems that use MapServer and/or OGC services emerged. Today projects like Chameleon, p.mapper, CartoWeb, MapBuilder, Mapbender and more recently Ka-Map and OpenLayers are thriving and have formed the basis for the development of a large number of FOSS4G projects.

With this success there is no compelling reason to develop client-side capabilities beyond what exists today. MapServer's role is clearly on the server-side in the production of maps and data in support any number of client environments. Recognizing this, and focusing resources, provides the key to keeping MapServer relevant as the Web-based GIS space becomes ever more crowded.

Even with the emergence of commodity mapping services like Google or Yahoo, it is suggested that MapServer still has an important role to play as a data provider. For example, it is trivial to author templates that expose data in various formats (e.g. KML or GeoRSS) that can be readily consumed by the large number of applications built atop of these services.

4.6.2 Better Cartography

MapServer 3.x focused primarily on abstracting data access and the development of MapScript, while MapServer 4.x focused largely on the implementation of the various OGC WxS standards. All the while the rendering internals of the software remained essentially the same. There were improvements made of course (e.g. path following labels) but most were for very specialized functionality such as arbitrarily rotated cross-hatching, and tended not to impact general rendering very much.

Version 5.0 of MapServer strived to improve rendering in ways that will benefit all users of the software. The most notable change was new support for the Anti-Grain Geometry (AGG) rendering library (http://www.antigrain.com/) in addition to continued support for GD. AGG holds the promise of high-quality, anti-aliased output that is simply not possible with GD, with limited performance reduction.

GD will continue to be a fixture within MapServer for the foreseeable future. For instance, the AGG support in MapServer 5.0 uses GD as a buffer manager. That is, AGG renders into a gdImagePtr buffer. This allows developers to make use of the GD image input/output functions, and they can fall back on GD rendering when it makes sense (or they do not have the time and/or resources to do it in AGG). Other renderers may hold promise in the future. For example, Cairo Graphics (http://cairographics.org/) supports multiple backends (image buffer, PDF, SVG) through a common API.

4.6.3 A Tile Engine

Tiled mapping applications have been available on the Web for years, from TerraServer to TopoZone. However, with the debut of the "slippy" interface popularized by Google, tiled rendering is back in vogue. Not surprisingly, MapServer is being utilized by several of the OS tiling packages, most notably Ka-Map and TileCache.

For years MapServer has been optimized to render individual maps as efficiently as possible. This is particularly apparent with label placement. MapServer generally places labels AFTER features have been clipped to the current view port. So, for example, if a lake polygon is only partially within a map only the portion within the map is used to place a label. This has the advantage of cramming as many labels as possible into the label cache for possible placement. However, in the case of a tiled (or seamless) map it is easily possible to end up with multiple labels for the same feature where it spans tile boundaries. Another example would be a thick line whose feature falls outside the clipping rectangle for a given map. Fortunately this problem was mostly addressed by enlarging the clipping rectangle based on the size of the largest symbol being used. However, this solution isn't perfect in all cases (e.g. when binding symbol size to an attribute).

The tiling engines have devised mechanisms to work around these shortcomings, such as the metatile approach employed by Ka-Map, but it would be advantageous if MapServer could operate in a "tiled" mode. MapServer 5.x will introduce several features specific to tiled image production. One is the ability to compute label points before clipping which should provide stable label placement regardless of the extent of a given tile. Another is the option to force images through a pre-defined 8-bit color palette, which will allow for the production of maps amenable to compression while maintaining consistent colours from tile to tile.

4.6.4 High-Performance Query Support

MapServer is now a relatively mature FOSS4G project and remnants of earlier versions can still be found. One important place where this applies is the use of two-stage querying. The first version of the software used shapefiles exclusively for input. In an effort to manage memory conservatively, feature identifiers (each element in a shapefile has a unique numeric identifier built into the format for fast access) are cached in a result set rather than storing entire features. This meant very large result sets could be handled without running out of memory. The feature identifier is then used to retrieve features quickly at presentation time. Simplicity in design and implementation was the trade-off for accessing features in two passes. Since queries tended to be an end-result for a user (as opposed to making a map) it was not as essential for these to be blazingly fast. The addition of non-shapefile data sources complicates matters since many don't have a unique numeric key (or require special setup to create this), or alternatively have high connection overhead that makes the second pass especially penal.

It is hoped that this limitation can be addressed sometime early in the 5.x version series either through a feature caching mechanism or by moving more of the query functionality off to the individual data source.

4.6.5 Standards Compliance

Support of OGC service specifications has been a boon to FOSS4G projects and small commercial providers alike. The developer community needs to continue to invest resources in keeping MapServer support as current as reasonable. As new standards or de facto standards emerge (e.g. KML, GeoJSON or GeoRSS), MapServer should be an early adopter whenever possible, independent of the potential costs this may incur. In this regard, a project like MapServer can be a great sandbox.

Moreover, the developer community needs (and has) MapServer representation in the various standards processes. Developers working on implementations can provide a nice balance to the somewhat academic nature of specification development.

4.6.6 Threaded Application Environments (C#, Java)

One of the big challenges with MapServer is taking what was originally written to run as a CGI process and make it viable as a long-running process. The code base has undergone a great deal of modernization in this regard. Global variables are avoided at all costs, memory leaks are at a minimum and locking is used to protect sections of code that are not thread-safe. Still, MapServer relies on a large number of supporting libraries and it is not always clear where they stand on these issues. Fortunately MapScript seems to be gaining popularity with both Java and .Net developers, thus progress is clearly being made.

4.6.7 XML

This is a topic that that comes up monthly on the mapserver-users mailing list. “Why aren’t configuration files in XML?” To be blunt, they just aren’t. To date no-one has had the resources or the energy to take that on, however it is likely that this will happen at some point in the near future. There are a number of benefits in following this approach, the biggest being the sheer number of tools that could factor into mapfile creation and management.

There are some more subtle advantages as well. For example, the current mapfile structure doesn’t differentiate between attributes and elements like XML does. So, for example, in a layer a name is just an element of a layer and as such it doesn’t make a very good primary key. Because it can show up anywhere in a layer definition, a developer doesn’t know what layer he/she is dealing with until the entire object is processed. This makes extending from external files or string buffers difficult. On the other hand, in XML the name can be an attribute of a layer element and when a layer is encountered it is clear exactly which object is being dealt with. Hence, it becomes easier to extend (and hence reuse) object definitions (or portions thereof).

4.7 Conclusion

By almost any measure MapServer has been a very successful FOSS4G project, and there doesn’t seem to be any clear reason why it can’t remain a fixture in this landscape for some time to come. This is particularly true if the project can remain focused on the qualities that made it useful in the first place.

The MapServer project owes a debt of gratitude to the large collection of other OS projects it so happily leverages, including Anti-grain Geometry, Apache, Autoconf, Automake, Bison, Flex, Freetype, GD, GDAL/OGR, GEOS, Make, Proj.4, libjpeg, libiconv, libpng, libtiff, libxpm, MySQL, Perl, PHP, PostgreSQL, PostGIS, Python, Ruby, Shapelib, Swig, Xerces, and of course all of their dependencies.

Chapter 5
The Geospatial Data Abstraction Library

Frank Warmerdam

Abstract This chapter presents an overview of the development and characteristics of the Geospatial Data Abstraction Library (GDAL), a widely used Open Source library for reading and writing a large variety of raster spatial data formats. The library has evolved substantially since its origins in 1998. It supports its own data model and application programming interface (API). From its initial single developer origins, GDAL has grown into a distributed project that has a relatively large number of contributing developers. The chapter discusses the origins of the project, its design philosophy, the data model, and directions for future development.

5.1 Introduction

GDAL is a C++ data access library for reading and writing a variety of spatial raster data formats. As a library it presents a single abstract data model to the calling application for all supported formats. It also comes with a variety of command line utilities for data translation and processing that are especially useful to end users, in contrast to the programmers who use the library. The library is split into a raster half and a vector half (OGR), each with its own data model and API. Both portions live in the same underlying library (i.e. GDAL14.DLL). For the purposes of this chapter, only the GDAL component of the library is discussed.

GDAL is widely used by in the open source geospatial world, including but not limited to packages such as MapServer (see Chap. 4), GRASS (see Chap. 9), QGIS, MapGuide (see Chap. 7), OSSIM, and OpenEV. It is also used to varying degrees by a variety of proprietary software products, including FME, ArcGIS, and Cadcorp SIS.

The project was launched in 1998 by the author and now includes 20 contributing developers. It is a foundation project of the Open Source Geospatial Foundation (OSGeo) and has cemented a place for itself as a tool of considerable utility both

Frank Warmerdam
Open Source Geospatial Foundation, Beaverton, OR, USA, e-mail: warmerdam@pobox.com

G.B. Hall, M.G. Leahy (eds.), *Open Source Approaches in Spatial Data Handling.*
Advances in Geographic Information Science 2,

to free and open source for geospatial (FOSS4G) application developers as well as end users who need access to translation utilities for working with spatial data in different formats.

5.2 Project Origins

The GDAL project had its origins in mid-1998 while the author was employed as a developer and designer of remote sensing software by PCI Geomatics (PCI). A determination to become an independent software developer, while protecting the investment of effort and intellectual property in developing code, was fundamental to two important decisions taken at this time.

It was initially decided that geospatial data translation would be an excellent focus for the development of a suite of tools for analysts and other users of spatial data. The wide range of spatial data file formats needed by software vendors meant there would potentially be no end to the work required to move data from one format to another. In most companies, writing file format translators was considered a low status responsibility, for which there would be little developer resistance to outsourcing. Personal experience at PCI led to the development of significant accumulated knowledge of various different file formats and their translation (i.e. non-glamorous programming). This work led to the role of primary architect of PCI's GeoGateway (GDB) data access library. Also while at PCI there was substantial personal effort invested in GeoTIFF standardization combined with release the open source (OS) shapelib library. Hence, there was a sound basis from which to embark on the development of GDAL.

It was decided from the outset to release GDAL as OS. This decision had its roots in use of GNU software over the previous decade as well as seeing the success of libtiff, libgeotiff and shapelib as OS. More importantly, following the OS path established a way of providing the assurance of unrestricted use of the code by clients, while preserving the personal ability to continue using, adapting and offering the code to new clients. This approach seemed to offer the best of all worlds, given the above objectives.

While at PCI, links had been forged by the author with other companies in the use of PCI's libraries. However, these efforts were often frustrated by the rightful concerns of other companies building their own proprietary systems with no assurance of future access for PCI, no indication of preferential pricing, and no voice in future development directions. Hence, leaving PCI and becoming an independent software developer under the guise of the OS movement was the outcome. From the outset, the design of GDAL was heavily influenced by the design of the GDB technology while working at PCI. The overall architecture was similar, while some decisions and approaches were made based on lessons learned with GDB. Hence, it is fair to say that GDAL was designed from its origins as a "better GeoGateway".

5.2.1 Lessons and the Influence of Libtiff

While GDB was in many ways the architectural model for GDAL, three existing OS libraries, namely libtiff, libgeotiff and shapelib, also each contributed in meaningful ways to the development of GDAL. Libtiff is an OS library developed in late 1980s, implementing the open TIFF image format. The library is still widely used in the spatial information technology industry as well as for other imaging applications.

The TIFF format, while fairly well specified in open documents, is nonetheless complex. Because of this there are relatively few full-featured implementations of the format. Instead most software vendors use libtiff, or one of a few proprietary libraries implementing TIFF. This lesson was hard learnt by early attempts to write TIFF reading and writing support for PCI software (in FORTRAN!) which met with only limited success. Hence, the first lesson learnt in the GDAL project was:

Lesson 1: Complex file formats are generally hard to implement completely, and it is much better to reuse a common implementation instead of re-inventing.

The libtiff library was licensed under a very liberal license akin to that used by MIT for licensing the widely used X Windows System (often referred to as the MIT/X-style license). Unlike the General Public License (GPL) or Lesser GPL (LGPL) licenses, the MIT license minimizes the "strings attached" to use of the software that is produced. Primarily it absolves the provider of any warranty, but it does not limit modification, redistribution or mixing with proprietary licensed software. This license was key to the use of libtiff by many proprietary software vendors. Hence, the second lesson suggests that:

Lesson 2: A permissive OS license (i.e. MIT/X) is important for gaining adoption of OS tools by proprietary software vendors.

Libtiff also modeled mechanisms where file system I/O, error reporting and even memory allocation were highly hookable. That is, applications could provide customized implementations of these interfaces with the outside world. To some degree these principles were also applied to GDAL. Hence, the third lesson learnt in the development of the GDAL project was:

Lesson 3: Libraries should make interfaces such as error handling, and file system input/output hookable for maximum application control.

5.2.2 Lessons and the Influence of Libgeotif

Libgeotiff is a library extension supporting the GeoTIFF format that brings the ability to use georeferencing and coordinate systems to TIFF files. Its availability formed the basis for the adoption of the GeoTIFF format. Beyond the value of the library and the format, GeoTIFF was a successful demonstration of bringing developers from a variety of commercial organizations together to cooperate. Hence, a further lesson suggests that:

Lesson 4: OS libraries can serve as a focus for co-operation among developers from many organizations.

5.2.3 Lessons and the Influence of Shapelib

Shapelib is a small library for reading and writing ESRI shapefiles that was first released by PCI. Technically, shapelib was not particularly sophisticated in its first release, and it represented only about a week's worth of development effort. However, it filled a valuable role at the time of its development, it was easy to use and, hence it proved to be very popular. In fact, this library is perhaps better known in the spatial data world than any other contribution made by the author, both before and after its publication. Use of Shapelib served to generate contract work, and besides its general utility it has continued to provide a useful revenue stream. Hence, the final lesson learned from this process is:

Lesson 5: Releasing a popular open library or component can be a powerful reputation builder, and a real marketing differentiator for a consultant or independent software developer.

5.3 Design Philosophy

As noted above, GDAL had its origins in the need for efficient and flexible file format translation that was extendable to the introduction of new formats and did not exclude commonly encountered spatial data formats for any specific design decision. Hence, the design aim from the outset was to achieve ubiquitous coverage through development of a unified abstract data model and adherence to standards under consideration or released by organizations such as the Open Geospatial Consortium (OGC). These design principles and others are discussed in the following sections.

5.3.1 A Unified Abstract Data Model

An important lesson carried over from GDB experience at PCI was to have a well defined abstract data model for implementation of GDAL. For GDAL, this is an abstract model of what a raster dataset may comprise, namely pixel data types, a metadata model, colour models, geo-referencing mechanisms, and so on.

While not all data format drivers implement all aspects of the abstract data model, the data model itself gives applications a generalized virtual model of datasets into which all formats fit. The data model is described in more detail in Sect. 5.4.

5.3.2 Align with the Open Geospatial Consortium

Where practical, GDAL seeks to be aligned with OCG standards. Because OGC does not specify C++ API standards, this generally did not take the form of implementing OGC specifications directly, but rather by closely aligning with the following OGC data models and data types:

OGC Simple Features Geometry: The OGC Simple Features specifications, such as Simple Features for SQL described a model for geographic information systems (GIS) vector geometries as well as well known text (WKT) and well known binary (WKB) encodings for these geometries.
OGC Spatial Reference Systems: The OGC Simple Features specifications also described a WKT mechanism for describing coordinate systems (projections). This representation was adopted by GDAL as its internal representation for coordinate systems. This standard is similar to the ESRI projection engine strings found in shapefile .prj files.
Grid Coverages: The OGC Grid Coverage specification defined a model for raster datasets (grid coverages). To some extent it was used as a model for aspects of the GDAL API. The GetCategoryNames(), GetOffset(), GetScale() and GetColorInterpretation() methods can be traced to the OGC coverage specification.

5.3.3 Zero Configuration – Point and Open

A further design lesson carried over from the GDB experience was to try and ensure that the opening of a dataset is as easy as possible. In particular, a user should not be expected to know the format of a file in advance or to need to set options before opening a dataset. It should be as easy as selecting a file and seeing if it is recognized by the software.

Within GDAL this is implemented by having a list of format drivers which are each (in turn) given a chance to try and open a dataset. The first that succeeds is used for the file. Individual drivers may recognize their formats based on recognized strings in file headers, extensions or in some cases special prefixes embedded in the dataset names. However, the key point is that a user should normally be able to use "File>Open" from the menu on any file without a priori knowledge about the files that are being opened.

5.3.4 Operating System Portability

GDAL aims to be portable to the Microsoft Windows Win32 operating systems (Windows 95 and all subsequent versions) as well as any POSIX-compliant system, including Linux and Unix operating systems. One fairly substantial omission is that no serious attempt has been made to support the pre-MacOS X operating system, although over time this is becoming increasingly less important.

From the beginning, portability to 64 bit operating systems and systems such as Solaris was also taken seriously. However, a conscious decision was made never to target relatively ancient or obsolete operating environments like 16 bit DOS and VMS. Portability was primarily aimed at currently popular operating systems, and at future proofing the library. The lack of MacOS 9x support was occasionally a problem but the decision to ignore it in favour of POSIX operating systems worked out well when Apple moved to a Unix-based operating system for the latest generations of their systems.

5.3.5 Focus on the Library

The focus for GDAL development has always been on its use as a library by other programmers, and the development of ancillary utilities beyond this basic objective has never been more than an afterthought. However, the project also produces command line utilities for doing translations and other operations with the library. Many of these utilities evolved from origins as test programs for the library. Hence, the command line options are often somewhat inconsistent and their documentation is relatively weak. At one time, work was started on a Qt-based demonstration viewer application as part of GDAL but this effort was terminated and discarded as a distraction.

Despite the relative weakness of the GDAL utilities, its focus as a library is based on the assumption that improvements to the core functions will provide the greatest general benefit since they help all users. That is, all users that work with client applications such as GRASS or FME, scripting language bindings such as those for Python, or the utility programs provided along with GDAL, will benefit from enhancements to the core library rather than enhancements to the development of peripheral utility functions.

5.3.6 Minimize Performance Overhead

A goal of GDAL is that applications should be able to read imagery with hardly any performance penalty relative to direct access. This goal exists in order to avoid application developers feeling the need to have their own custom "high performance" raster API for an internal working format. In practice this objective is accomplished by exposing the underlying format's optimal block size and ensuring that a RasterIO() request for a full block in the native data type is accomplished with a minimum of overhead. Generally, just two raw data copies are used, one is placed into the cache block and then one copy is placed from there to the application buffer.

To avoid even this double copy, the API also includes the ReadBlock() method to call the driver block reading function directly without going through the block cache. While this continues to work, its use is generally discouraged for all but the most performance-focused applications.

5.3.7 Easy Windowing and Type Conversion

The GDAL RasterIO() method provides the ability to pull an arbitrary window of raster data with up or down sampling and with pixel data type conversion, all on the fly. This makes it much easier to write applications that support any organization and pixel data type of raster file.

This can be seen as an anti-lesson from the libtiff library. The libtiff library exposes the underlying data organization but does not provide support for windowing, or data type conversion. This means that applications wanting to support the full family of TIFF files (tiled versus striped, different pixel data types, pixel interleaved versus band interleaved) end up having to implement many different cases in application code. GDAL takes the opposite approach, insulating the application from having to handle these cases explicitly while still supporting optimized access when needed. The down sampling support also makes transparent use of overviews if they are available.

5.3.8 Support Non-file Datasets

Not all image or feature sources are files in the file system. Hence, GDAL avoids the assumption that it can only operate on files. This applies to raster sources, where, in some cases, datasets might be a reference to an RDBM system, or to a remote server.

While GDAL supports such non-file datasets fairly well, the same cannot be said for some applications which still make it very difficult to enter non-file dataset names.

5.3.9 Easy to Implement Simple Drivers

It should be relatively easy to write a driver for a simple file format. This means there should be a minimum of methods that are required to be implemented for new drivers, with reasonable fallback implementations for missing methods. For GDAL, simple read-only drivers with no geo-referencing might only be needed to implement the Open() method to create and populate the data source, and the IReadBlock() method to read the imagery for the raster band.

Of course, the more capabilities a driver supports, the more involved the driver implementation is, and developing full read/write drivers with geo-referencing, various optimizations, and a variety of types of metadata can be fairly involved.

5.4 GDAL Data Model

The GDAL data model comprises a number of components each of which subscribes to the basic design philosophy of the libaray as outlined above. These components are discussed in the following sections.

5.4.1 GDALDataset

A dataset (represented by the GDALDataset class) is an assembly of related raster bands and some information that is common to all of them. The dataset has a concept of the raster size (in pixels and lines) that applies to all of the bands. The dataset is also responsible for the georeferencing transform and coordinate system definition of all bands. The dataset itself can also have associated metadata stored as a list of name/valuepairs in string form, as specified below.

The GDAL dataset and raster band data model is loosely based on the OGC Grid Coverages specification. The reason for this loose relationship is discussed in Sect. 5.6.3.

5.4.2 Coordinate Systems

GDAL dataset coordinate systems are represented as OGC WKT strings. These can contain:

- Overall coordinate system name.
- Geographic coordinate system name.
- Geographic coordinate system name.
- Datum identifier.
- Ellipsoid name, semi-major axis, and inverse flattening.
- Prime meridian name and offset from Greenwich.
- Projection method type (e.g., Transverse Mercator).
- List of projection parameters (e.g., central_meridian).
- Unit's name, and conversion factor to meters or radians.
- Names and ordering for the axes.
- Codes for most of the above in terms of predefined coordinate systems from authorities such as the European Petroleum Survey Group (EPSG) (http://www.epsg.org/).

An empty coordinate system string indicates nothing is known about the georeferencing coordinate system that is being used.

5.4.3 Affine GeoTransform

GDAL datasets have two ways of describing the relationship between raster positions (in pixel/line coordinates) and georeferenced coordinates. The most commonly used method is the affine transform. This consists of six coefficients returned by GDALDataset::GetGeoTransform() which map pixel/line coordinates into georeferenced space using the following relationship:

$$\text{Xgeo} = \text{GT}(0) + \text{Xpixel}^{*}\text{GT}(1) + \text{Yline}^{*}\text{GT}(2)$$
$$\text{Ygeo} = \text{GT}(3) + \text{Xpixel}^{*}\text{GT}(4) + \text{Yline}^{*}\text{GT}(5)$$

In the case of "north up" images, the GT(2) and GT(4) coefficients are zero, the GT(1) is pixel width, and GT(5) is pixel height. The (GT(0),GT(3)) position is the top left corner of the top left pixel of the raster.

Note that the pixel/line coordinates in the above are from (0.0,0.0) at the top left corner of the top left pixel to (width_in_pixels,height_in_pixels) at the bottom right corner of the bottom right pixel. For example, the pixel/line location of the centre of the top left pixel would therefore be (0.5,0.5).

5.4.4 Geographic Control Points

The second method that GDAL has for describing the geographic reference of a raster dataset uses geographic control points (GCP). With this method, a dataset will have a set of GCP relating one or more positions on the raster to georeferenced coordinates. All GCP share a georeferencing coordinate system (returned by GDALDataset::GetGCPProjection()). Each GCP (represented as the GDAL_GCP class) contains the following:

```
char *pszId
char *pszInfo
double dfGCPPixel
double dfGCPLine
double dfGCPX
double dfGCPY
double dfGCPZ
```

The pszId string is intended to be a unique (and often, but not always numerical) identifier for the GCP within the set of GCPs on a dataset. The pszInfo is usually an empty string, but can contain any user-defined text associated with the GCP. Potentially this can also contain machine parsable information on GCP status, though this is not implemented as yet. The (Pixel,Line) position is the GCP location on the raster. The (X,Y,Z) position is the associated georeferenced location (with the Z often being zero).

The GDAL data model does not imply a transformation mechanism that must be generated from the GCP as this is left to the application. However 1st to 5th order polynomials are common.

Normally a dataset will contain either an affine geotransform, GCP, or neither. It is uncommon to have both, and it has not as yet been determined which method is more authoritative.

5.4.5 Metadata

GDAL metadata are auxiliary format- and application-specific textual data kept as a list of name/value pairs. The names are required to be well behaved tokens (no spaces, or odd characters). The values can be of any length, and contain anything except an embedded null (ASCII zero).

Some formats will support generic (user-defined) metadata, while other format drivers will map specific format fields to metadata names. For instance the TIFF driver returns a few information tags as metadata including the date/time field which is returned as:

TIFFTAG_DATETIME=1999:05:11 11:29:56

Metadata are split into named groups called domains, with the default domain having no name (NULL or ""). Some specific domains exist for special purposes. Note that currently there is no way to enumerate all the domains available for a given object, but applications can "test" for any domains they know how to interpret. One example is the SUBDATASETS domain, described in the following section.

5.4.6 SUBDATASETS Domain

The SUBDATASETS domain holds a list of child datasets. Normally this is used to provide pointers to a list of images stored within a single multi-image file (such as HDF or NITF). For instance, an NITF with four images might have the following subdataset list:

SUBDATASET_1_NAME=NITF_IM:0:multi_1b.ntf
SUBDATASET_1_DESC=Image 1 of multi_1b.ntf
SUBDATASET_2_NAME=NITF_IM:1:multi_1b.ntf
SUBDATASET_2_DESC=Image 2 of multi_1b.ntf
SUBDATASET_3_NAME=NITF_IM:2:multi_1b.ntf
SUBDATASET_3_DESC=Image 3 of multi_1b.ntf
SUBDATASET_4_NAME=NITF_IM:3:multi_1b.ntf
SUBDATASET_4_DESC=Image 4 of multi_1b.ntf
SUBDATASET_5_NAME=NITF_IM:4:multi_1b.ntf
SUBDATASET_5_DESC=Image 5 of multi_1b.ntf

The values of the tokens ending with "_NAME" are strings that can be passed to GDALOpen() to access the files. The values of the tokens ending with "_DESC" are intended to be more user friendly strings that can be displayed to the user in a selection list.

5.4.7 Raster Band

A raster band is represented in GDAL with the GDALRasterBand class. It represents a single raster band, channel, or layer. It does not necessarily represent a whole image. For instance, a 24 bit RGB image would normally be represented as a dataset with three bands, one for red, one for green and one for blue. A raster band has the following properties:

- Width and height in pixels and lines. This is the same as that defined for the dataset, if it is a full resolution band.
- Datatype (GDALDataType). One of Byte, UInt16, Int16, UInt32, Int32, Float32, Float64, and the complex types CInt16, CInt32, CFloat32, and CFloat64.
- Block size. This is a preferred (efficient) access chunk size. For tiled images this will be one tile. For scanline-oriented images this will normally be one scanline.
- List of name/value pair metadata in the same format as the dataset, but of information that is potentially specific to this band.
- Optional description string.
- Optional list of category names (effectively class names in a thematic image).
- Optional minimum and maximum value.
- Optional offset and scale for transforming raster values into meaningful values (e.g., translate height to metres)
- Optional raster unit name. For instance, this might indicate linear units for elevation data.
- Colour interpretation for the band. This is one of undefined, gray, palette index, red, green, blue, alpha, hue, saturation lightness, cyan, magenta, yellow and black.
- A colour table, described in more detail in the following section.
- Knowledge of reduced resolution overviews (pyramids), if available.

5.4.8 Colour Table

A colour table consists of zero or more colour entries described in C code by a structure containing:

```
short c1; /- gray, red, cyan or hue -/
short c2; /- green, magenta, or lightness -/
short c3; /- blue, yellow, or saturation -/
short c4; /- alpha or blackband -/
```

The colour table also has a palette interpretation value (GDALPaletteInterp) which is one of the subsequent values, and indicates how the c1/c2/c3/c4 values of a colour entry should be interpreted. The values are gray, rgb, cmyk and hls.

To associate a colour with a raster pixel, the pixel value is used as a subscript into the colour table. That means that the colours are always applied starting at zero and ascending. There is no provision for indicating a pre-scaling mechanism before looking up in the colour table.

5.4.9 Overviews

A band may have zero or more overviews. Each overview is represented as a "free standing" GDALRasterBand. The size (in pixels and lines) of the overview will be different than the underlying raster, but the geographic region referenced by overviews is the same as the full resolution band. The overviews are used to display reduced resolution overviews more quickly than could be done by reading the full resolution data and down sampling it.

5.5 What Worked? What Didn't?

With the GDAL data model described above, it is pertinent to consider the relative successes and failures of the project. While it is fair to say that more aspects of the library development were successful than were not, there are aspects of the lessons, mentioned earlier, learned along the way that have served to direct the development effort. These aspects of GDAL development are now discussed.

5.5.1 Contributions from Developers

The GDAL project, like most OS projects, has welcomed contributions from a variety of developers since its inception. At the time of writing GDAL has 28 authorized contributors (i.e., developers with the rights to make changes directly to the software repository). Of these, perhaps half are reasonably active, with some only taking responsibility for a limited area of the code, such as a single driver or language binding. However, all 28 have contributed at least some code to the project. In addition, significant amounts of code have been contributed as patches from many additional contributors, including whole drivers, features, bug fixes, and documentation. The main downside to accepting contributions from a large number of developers is the difficulty of maintaining consistency in the approach and quality control across all participants and the code base. However, the benefits have certainly outweighed the costs of this variability.

Arguably, the formation of a Project Steering Committee (PSC), and the guarantees of community equality that are implicit in the project becoming part of OSGeo have helped accelerate the number of contributors. However, it is hard to be completely confident of this. As part of the launch of the PSC, a Request For Commands (RFC) process was introduced to allow anyone to bring forward a proposed and carefully documented architectural change or addition. With this process, and the PSC making the final decision to adopt a proposal, there is now a fair opportunity for any developer to propose and then implement improvements. This hopefully will avoid some of the frustration of the past when suggestions might have been too readily or arbitrarily dismissed by the author. Hence, developer contributions and the corresponding efforts to build a developer community have proven to be a success for both GDAL and OGR.

5.5.2 C/C++ Implementation Language

GDAL is implemented in C++ (in some cases with C libraries). At the time of the project launch a case could have been made for implementing it in Java. Nine years later, a case might instead be made for writing it in a language like C# (.NET). However, one of the key original goals of the project was to achieve ubiquity. To a large extent this goal has been met, and part of the reason for this is that GDAL is portable and has relatively light requirements in terms of the provided environment. While Java is also portable, it comes at a large cost in terms of its dependence on the presence of a Java Virtual machine (JVM) that makes it hard to embed in applications without adding complex new requirements. Similarly C# (.NET) would add dependencies with its complex and large runtime environment and this raises questions about portability beyond the native windows C# environment.

In practical terms, many of the proprietary software vendors that use GDAL and are responsible for financially supporting development of the project are building C/C++ applications. These vendors would not have been willing to incorporate libraries in other languages that would complicate their build and run time environments.

C/C++ is the one implementation language that is fairly easy to wrap using a simplified wrapper and interface generator (SWIG) approach to make the library capabilities available in other languages like Python, Perl, C#, and Java. To a significant degree, C/C++ is still the language of choice for library technologies that can be accessed from many different languages and environments.

One downside of implementation in C/C++ is that it is easy to build fragile, buggy applications due to the "bare metal" methods of accessing and managing memory. The other downside of C++, in particular (compared to C), is the fragility of the application binary interface (ABI). If two versions of a library are ABI compatible, it is possible to replace one with the other (as a DLL or shared library) without breaking the application. However, applications written to the GDAL C++ API are generally only able to use other builds of GDAL of exactly the same version, and that are built with essentially the same compiler. This is because C++ ABIs are very sensitive to any change in the size and layout of objects, and because C++ compilers use subtly different ways of translating ("mangling" in compiler terminology) class and method names into object code names. As well, C++ depends on a quite complex and rapidly evolving runtime environment.

For this reason, the GDAL ABI is very fragile. The main workaround for this has been to wrap the C++ API with a C API. In the C API, all object pointers become opaque handles, so object size and layout changes between versions don't matter. C compilers generally use standardized (and very simple) name mangling, and depend on a stable and relatively simple runtime API. So the C API is quite stable, and it is now project practice to encourage external applications to use the C API specifically to ensure better cross version ABI compatibility.

Thus, implementation in C/C++ has generally been a success, while initially choosing to present a C++ API to applications was a failure because of ABI fragility.

5.5.3 OGC Orientation

As mentioned earlier, GDAL attempts to build around the models and definitions of appropriate OGC specifications, where available. In particular, the raster API is somewhat organized around the OGC Grid Coverages specification. The OGRSpatialReferences uses the OGC WKT representation for spatial reference systems as its internal model.

In practice, use of the Grid Coverages specification has not had any apparent benefit. Some of the methods, colour models, and attributes that were added because they were in the grid coverages specification have essentially never been used. These are therefore "cruft" (i.e., useless or redundant) and at times confusing additions to the library. It likely would have been better to include only features of the Grid Coverages specification when there was a clear requirement to do so.

In contrast, organization of the OGRSpatialReference class and coordinate system handling around the OGC WKT coordinate system representation have proven useful. It is a well conceived data model that was independently adopted by other software packages which were subsequently easier to integrate with GDAL (e.g., Cadcorp SIS, and MapGuide). It has also made development of file format drivers for formats like PostGIS, Oracle Spatial, ESRI Shapefiles and ESRI Personal Geodatabases easier because they are built around variations of OGC WKT coordinate systems. The downside is that the original OGC WKT spatial reference system specification was not as complete and specific as it should have been, and this has contributed to a proliferation of similar but subtly different implementations.

Overall, OGC-orientation has been a big success with regard to geometry and spatial references, and a mild failure with regard to grid coverages.

5.5.4 MIT/X License

As noted earlier, the MIT/X license is a very simple OS license that was first used for the Massachusetts Institute of Technology distribution of the X Window System. The terms essentially allow the licensee to copy, modify, and redistribute the licensed code as long as they don't change the copyright notices on the code itself, and they accept a disclaimer of warrantee. In addition, it makes no attempt for users of the software to provide the source code, or their own modifications or source code for the rest of their software. These sorts of conditions are found in the GPL and LGPL licenses used for OS projects including GRASS and Linux. Such terms are sometimes referred to as "infectious" or "virial" because they try to force redistributors of the software to retain the freeness of the software, even expanding it to all the software linked with a specific component.

It was a deliberate decision of the GDAL project not to use a license like the LGPL or GPL. These licenses are quite confusing, and can make it essentially impossible for a proprietary software vendor to use OS libraries under the licenses without having to give up control of their own software. Adoption by proprietary software vendors is part of the GDAL goal of being ubiquitous, so the most appropriate license available was chosen. This approach has been a success and the MIT/X licensing of GDAL has not been a barrier to the widespread adoption of the software.

5.6 Future Directions

After eight years, the GDAL project has established a stable role for itself within the proprietary GIS and FOSS4G communities. The future will bring new formats and further refinements to the support for existing formats. However, there are some specific areas that will likely receive further work, though addressing them will depend on community demand, and at best will take several years to complete. Some of the refinements either under consideration or already underway are discussed in the following sections.

5.6.1 Thread Safety

Workstations with multiple central processing units (CPUs) or with CPUs that contain multiple processing cores are becoming common and applications are increasingly employing multi-threaded programming techniques to take advantage of them. Making a library thread-safe means ensuring that the library can be executed from multiple points within the same program. A typical strategy to deal with this is to put thread locks around critical sections of code that update shared data structures.

Some work has already been done in GDAL focused on making read-only raster access thread-safe. However, much work still remains to be done. There are several levels of thread safety that can be achieved. The current objective is to aim for applications to be able to utilize distinct datasets in different threads at the same time. A harder objective would be to allow a single dataset to be used in multiple threads at the same time, but this is almost impossible without having to add locking in every GDAL method.

The main complicating factor in implementing thread safety in GDAL is the use of shared static variables. The central case of this is the shared block cache. This is a list of raster blocks stored in memory in a single data structure shared between all datasets. Appropriate locks have been introduced into this core to handle the read-only case, but it turns out that having writable datasets is substantially harder, as reading a block on one dataset may require flushing a block to disk that belongs to a dataset that a different thread is operating on.

Thread safety also requires careful review of code for use of static variables in convenience and other mechanisms (such as remembering error messages). Only a few GDAL drivers have been carefully vetted in this regard, though the GDAL core is in good condition. Applications such as MapGuide, ArcGIS, and MapServer are the primary drivers of interest in GDAL thread-safety. It is also a significant issue for language bindings such as Java and .NET, which typically operate in a multi-threaded fashion.

5.6.2 Internationalization

GDAL uses the eight bit character data type for essentially all text. However, in the real world there are many different character sets. Even for western character sets there exist many different "code pages" that contain different accented or other special characters. To some extent, the project has been able to ignore these distinctions with western character sets. However, to support completely distinct character sets with more than 256 glyphs, such as Chinese or Japanese character sets, it is critical to handle different character sets explicitly.

Many applications approach this by switching to use of "wide characters", a 16 or 32 bit character type which can support all characters used worldwide via a master dictionary called UNICODE. Introducing wide characters into use in GDAL would be a substantial complication, and would also be very disruptive for existing applications using the library. Thus, the tentative plan is to switch to a mechanism called UTF-8, which is an encoding of text into eight bit characters, with some characters requiring multiple characters to be fully represented. The benefit of this is that most of the application can pass around UTF-8 strings using the same methods already in use with a minimum of disruption. However, some parsing code will need to be aware of the specific character boundaries. Specific file format drivers, especially OGR drivers, will have to be more aware of the codepage being used for feature attributes to set up the string values accordingly. While these issues only apply to metadata and filenames for GDAL, they are much more pressing for the OGR component of the overall library, where it is quite common for feature attributes to be in a wide variety of languages.

In addition to changes within GDAL, it is also likely that the library will become dependent on the iconv library, a library for transliteration of text between different character sets. This is already being used successfully in the MapServer project as well as many other projects.

Another aspect of internationalization is translating the GDAL documentation, error messages, and driver documentation into different languages. While this may occur at some point in the future, it is not on the near horizon.

5.6.3 Improved Java, .Net Bindings

As noted earlier, GDAL supports bindings in a number of languages including Python, Perl, Ruby, C# (.NET) and Java. This is accomplished using the SWIG utility to generate in a fairly automated fashion language specific "jackets" for all the functions in GDAL. This is working reasonably well now, and especially the Python bindings have a substantial user base.

The C# and Java bindings are somewhat newer, but due to issues with SWIG and the languages themselves they are more difficult to implement. The Java and C#/.NET languages both have huge programmer communities and improving the language bindings will be important to provide support for developers working from within these communities. This support includes fixing bugs in the bindings, filling in omitted functions, adding "syntactic sugar" to make the bindings easier and more natural in the target language, making building easier, and, most of all, improving language specific documentation.

5.6.4 GDAL/OGR Grand Unification

Conceptually GDAL and OGR are distinct libraries that happen to share a few common components. In fact, they have quite distinct approaches to satisfying different needs. For instance GDAL has a concept of metadata, which is completely missing in OGR. OGR uses TestCapability() methods on many objects so that applications can determine what a particular format does or does not implement. When originally written, this seemed reasonable since it was perceived that GDAL and OGR would be used separately by distinct applications. However, this has not actually been the case. In practice many applications use both parts, and the very distinct approaches to some problems can cause difficulties and at the very least a cognitive dissonance.

GDAL/OGR "grand unification" is about merging the two halves to the extent that is practical. This might include:

- A single unified driver model – one driver class replacing GDALDriver and OGRSFDriver.
- Merging GDALDataset and OGRDataSource into a common "fat" dataset class, or deriving them from a common base class.
- A unified GDALOpen() for raster and vector files.
- Use of GDALMajorObject as a base class providing metadata services to OGR-Layer.
- Incorporation of TestCapability in GDALDataset and GDALRasterBand for testing capabilities.
- Grand unification still forsees the OGRLayer and GDALRasterBand being distinct.

This unification would result in substantial API breakage, and the resulting version would likely be known as GDAL 2.0. All reasonable efforts would be made to minimize loss of backwards compatibility. This would likely be more achievable on the GDAL side, while OGR applications would require at least some source changes. Hence, the status quo with enhancements to the two libraries is likely to remain in place in the immediate future. However, if a convincing case can be made for expending the design and implementation effort that would be required to merge the libraries into a unified whole then more attention may be placed on this venture.

In their current forms, the libraries are both widely used in both the FOSS4G and proprietary geospatial communities. The initial goal noted at the outset of this chapter of allowing the author to operate as an independent software developer while retaining the IP embodied in tools produced has been achieved. Recent consolidation of the FOSS4G community through the non-profit organization OSGeo serves as important point of reference for like-minded developers that should allow projects such as GDAL/OGR to build upon their initial successes within a growing body of contributors.

Chapter 6
Open Source Databases and Their Spatial Extensions

Rongguo Chen and Jiong Xie

Abstract This chapter discusses Open Source databases including two modes of spatial data management, two different key standards for spatial databases, the leading Open Source databases including MySQL, PostgreSQL, FireBird, Ingres and MaxDB, and summarises extending methods and the use of PostGIS and MySQL.

6.1 Introduction

This chapter is devoted to discussing several open source databases (OSDBs) and their extensions for spatial data handling. It starts with a general introduction of OSDBs focusing on MySQL, PostgreSQL, FireBird, Ingres and MaxDB. A function and source code comparison of them are given. Thereafter, two approaches to develop spatially enabled OSDBs are presented, namely a mid-ware spatial database engine (SDE) solution and a database management system (DBMS) with a spatial extension, or a spatial DBMS. In Sect. 6.4, three kinds of derived standards for OSDB spatial extension development are introduced, namely ISO SQL/MM, ISO/TC 211 (Geographic information/Geomatics) and OGC SF-SQL. In Sect. 6.5, the extending method of PostgreSQL is illustrated together with the use of PostGIS and MySQL. Oracle Spatial, one of the most widely used commercial spatial databases, is compared with the OS alternatives is also given. Likely future developments of OSDBs are discussed in the last section of the chapter.

Rongguo Chen
State Key Laboratory of Resources and Environmental Information System,
Chinese Academy of Sciences, Beijing, China, e-mail: chenrg@lreis.ac.cn

Jiong Xie
State Key Laboratory of Resources and Environmental Information System,
Chinese Academy of Sciences, Beijing, China, e-mail: xiej@lreis.ac.cn

G.B. Hall, M.G. Leahy (eds.), *Open Source Approaches in Spatial Data Handling.*

6.2 A Review of Open Source Databases

OSDBs have received a great deal of attention for some time. MySQL, PostgreSQL, Ingres, Firebird and MaxDB are the most popular, and among these MySQL, which is dual licensed, is most extensive. MySQL is available on more than twenty different platforms, including major Linux distributions, Mac OS X, UNIX and Microsoft Windows. Its architecture, based on threads in Windows or processes in UNIX, makes it extremely fast and easy to customise. It has all of the major characteristics of commercial enterprise-level databases including full transaction-safe integration, atomicity, consistency, isolation, durability with ACID-compliance, with full commit, rollback, crash recovery and row level locking capabilities. Versions after V4.1 implement a subset of structured query language (SQL) with the geometry types environment proposed by the Open Geospatial Consortium (OGC). A geometry-valued SQL column is implemented to support the simple feature data (SFD) model.

PostgreSQL is a highly scalable, SQL-compliant, open source relational database system. It runs on all major operating systems, including Linux, UNIX and Windows. It has a proven architecture that has earned it a strong reputation for reliability, data integrity, and correctness. Like MySQL, it is also fully ACID-compliant, and has full support for foreign keys, joins, views, triggers, stored procedures, storage binary large objects, programming interface for C/C++, Java, .Net, Perl, Python, Ruby, Tcl, and ODBC. PostgreSQL also has numerous sophisticated features including multi-version concurrency control (MVCC), point in time recovery, tablespaces, asynchronous replication, nested transactions, online/hot backups, a sophisticated query planner/optimizer, and write ahead logging for fault tolerance. It allows geometric data types (i.e. point, line, lseg, box, path, polygon and circle) to represent internally two-dimensional spatial objects. However the internal spatial objects have had limited applications until recently because of its non-standardization. PostGIS, as a spatial extension of PostgreSQL, conforms with the OGC and ISO SQL/MM standards, which make it possible to use PostgreSQL for modern spatial application development.

Ingres is an enterprise-class OSDB, which is dual licensed. It is a reliable, high-performance relational database solution that offers scalability, integration, and flexibility (Ingres 2006). Ingres scales from a laptop to workstation clusters in a single version and has the lowest total cost of ownership in the commercial enterprise-level database industry. Its enterprise-class features such as cluster solution, parallel query, distributed transactions are available under an OS license, but the support for B1 security, Enterprise Access, EDBC products are not open. Designed for "set-and-forget" operations on a compact footprint, Ingres can function as an unattended database for both enterprise and embedded deployment. Ingres is the ancestor of PostgreSQL, and it has an extension similar to PostgreSQL, called spatial object library supporting two-dimensional spatial objects, which is proprietary.

Firebird is derived from Borland InterBase 6.0 source code. It is open with many ANSI SQL-standard features and has no dual license (Borrie 2006). It can also run on multiple platforms including Linux, UNIX and Windows. The software has two main components, namely the database server and application interface. It provides excellent concurrency, high performance, and powerful language support for stored procedures and triggers. Firebird server has a very small "footprint" on the filesystem of a host server. A full installation, with all tools and documentation, needs less than 10 Mb.

MaxDB is a re-branded and enhanced version of SAP DB. It is a heavy-duty, systems application and products (SAP)-certified OSDB for online transaction processing (OLTP) and online analytic processsing (OLAP) usage which offers high reliability, availability, scalability and a very comprehensive feature set. It is targeted for large MySAP Business Suite environments and other applications that require maximum enterprise-level database functionality. In this regard it complements the MySQL database server (MaxDB 2005). Table 6.1 compares the main functions of the above OSDBs.

In summary, Ingres has been extensively developed more with relatively integrated functions, which can meet the requirements of the enterprise applications,

Table 6.1 Comparison of the main functions of OSDBs

		MaxDB 7.6	PostgreSQL 8.2	FireBird 2.0	MySQL 5.0	Ingres 2006
ACID		Support	Support	Support	Support (InnoDB table)	Support
Associated integrity		Support	Support	Support	Support (InnoDB table)	Support
DB transactions		Support	Support	Support	Support (InnoDB table)	Support
Unicode		Support	Support	Support	Support	Support
Index	R–/R+	No support	Support	No support	Only MyISAM	Support
	Hash	No support	Support	No support	Only InnoDB	Support
	GiST	No support	Support	No support	No support	No support
Temp table		Support	Support	No support	Support	Support
Materialized view		No support	No support	Only common views	No support	Ingres r4
Table partion	Range	No support	Support	No support	Support	Support
	Hash	No support	Support	No support	Support	Support
	List	No support	Support	No support	Support	Support
Distributed transactions		No support	No support	Support (2PC)	Support (XA)	Support
Cluster		Support	Support by add-on	No support	Support	Support
Spatial extension		No support	Support	No support	Support	Support

Table 6.2 Comparison of the source code of the main part of OSDBs

	MaxDB 7.6	PostgreSQL 8.2	FireBird 2.0	MySQL V5.0	Ingres2006
Files	1,203	1,081	913	2,353	5,696
Functions	4,692	9,506	7,075	30,994	22,470
Lines of code	287,792	374,124	584,431	890,415	1,440,326
Lines of comments	103,035	155,720	254,937	286,385	1,373,997
Ratio comments/code	0.36	0.42	0.44	0.32	0.95

while FireBird, with its smaller structure and better functions, can be used for middle or small database applications. MySQL and MaxDB have both scaleable and non-scalable applications. PostgreSQL is designed with an advanced architcure and has mature spatial extensions.

Besides the comparison of functions in Table 6.1, a tool named "Understand for C++" (Version 1.4) by Scientific Toolworks Inc. was used to check the source code of the main parts of the above mentioned OSDBs (all of the source code of MySQL was checked because it has several storage engines). The results of this analysis are shown in Table 6.2.

From this, it is clear that Ingres has more lines of code than any of the other products, while MySQL has slightly more functions. The ratio of comments to code is also the highest for Ingres. More comparisons of OSDBs are shown in Fabalabs (2005).

6.3 Spatial Extensions for OSDBs

Due to the features of spatial data such as location, non-structure (variable-length, such as the geometry field), spatial relationships, classification and encoding, it is possible to enhance their management by adding corresponding functions to a DBMS. Historically, solutions to the management of spatial data have been defined as a mixture of files and RDBMS, an entire RDBMS driven with a SDE, a spatial extension of RDBMS/object-relational database management system (ORDBMS), and other variants. Over time, the RDBMS with a SDE and the spatial extension of RDBMS/ORDBMS options have become recognised as the most popular solutions.

6.3.1 Spatial Database Engine (SDE)

An RDBMS with a SDE is a mid-ware solution that has been promoted by GIS vendors during the past decade. With this approach, the spatial data are put into the database by the SDE just as planting seeds through a channel. The engine picks up the data from the database when instructed by the user and transforms the data into the mode the user requires. Thus, operations and transactions of spatial data are processed out of the kernel of the database. The relational database is no more than a container of spatial data while the engine is just a channel through which the data pass. Products such as ArcSDE of the Environmental Systems Research Institute (ESRI) and SpatialWare of MapInfo represent this type of approach. They support general RDBMSs and can cross database platforms. The spatial data are generally stored and picked up in binary large object (BLOB) format, which is compatible with different types of databases and can be compactly linked with given GIS platforms.

This mid-ware solution has the additional advantages of using standard off-the-shelf components and reusing generic data management code (Breunig et al. 2003). There is a clear separation of responsibilities in that spatially-specific development can be performed and supported independently of the DBMS development. However, this approach may lose some efficiency due to the channel mapping and cross multi-database-platform mechanism. Moreover SQL extensions and data sharing and inter-operating are difficult because the data model is comparatively complicated. The more effort that is put into such a spatially-specific data management extension, the more difficult it becomes to change the system and to take advantage of DBMS improvements.

6.3.2 Spatially-enabled DBMS

Object-relational DBMSs (ORDBMS) are implemented by database vendors to manage spatial data. ORDBMS support the definitions of abstract data types including their operations and non-first normal form (N1NF) to organize non-structured data, including spatial data (note that a RDBMS with recent SQL features can also do this but not as well). The spatial data are stored and picked up as objects. The user can write types and functions/methods of spatial data with standard extended SQL in which new data types and functionality are integrated as closely as possible into the DBMS, and spatial data are integrated completely with other data. Traditional DBMS functionality like indexing, query optimization, and transaction management are supported in a seamless fashion for user-defined data types and functions (Breunig et al. 2003).

The database works directly at the command of the user instead of through a SDE broker, as described above, which makes it easier to implement spatial operations and generally results in good performance. Its disadvantage is that the architecture must be bound to a specific database platform. PostgreSQL/PostGIS, MySQL Spatial Extension, Ingres Spatial Object Library, Oracle Spatial, IBM DB2 Spatial Extender and Informix Spatial DataBlade are examples of this approach.

6.3.3 Building Blocks of Open Source Spatial Management Systems Database

There are basically two ways to develop spatially-enabled OSDBs. One is to implement a SDE for each OSDB, for example, SDE for MySQL and PostgreSQL. The other is to extend the OSDB's kernel to support spatial data based on industry standards. PostGIS and MySQL Spatial Extension are representatives of the second type.

6.4 Standards for Spatial Databases

There are three kinds of standard for spatial DB, namely the ISO SQL/MM, the standards defined by ISO/TC 211, and the OGC SFSQL specification. These standards are defined for different purposes (Ashworth et al. 1998), and there are differences in naming and terminology. The ISO/TC 211 Geographic information/Geomatics standards define components that may be assembled in an application schema to define a data product. The other two standards define generic types to be used in a SQL environment. The OGC has complementary specifications for the Common Object Request Broker Architecture (CORBA) and Common Object Model (COM) environments. ISO/TC 211 standards have a broader scope including 3D support and complexes, while OGC SFSQL and ISO SQL/MM address simple 2D elements. The OGC SFSQL and ISO SQL/MM have types to manage collections of points, curves, surface and a mixture of primitives. This facility is not apparent in the ISO/TC 211 standards, although they have the ability to compose complexes by reference.

6.4.1 OGC's Implementation Specification for Geographic Information

The purpose of the OGC's SFSQL is to define a SQL schema that supports storage, retrieval, query and update of feature collections via the SQL call-level interface. A geographic feature is an abstraction of a real world phenomenon associated with a location relative to the earth (OGC 2006a). It has both spatial and non-spatial attributes. Spatial attributes are geometry valued, while non-spatial attributes are classical SQL data type valued.

6.4.1.1 OGC Geometry Type Hierarchy

The OGC SQL Geometry Types are organized into a type hierarchy as shown in Fig. 6.1. The root type, named Geometry, has subtypes for Point, Curve, Surface

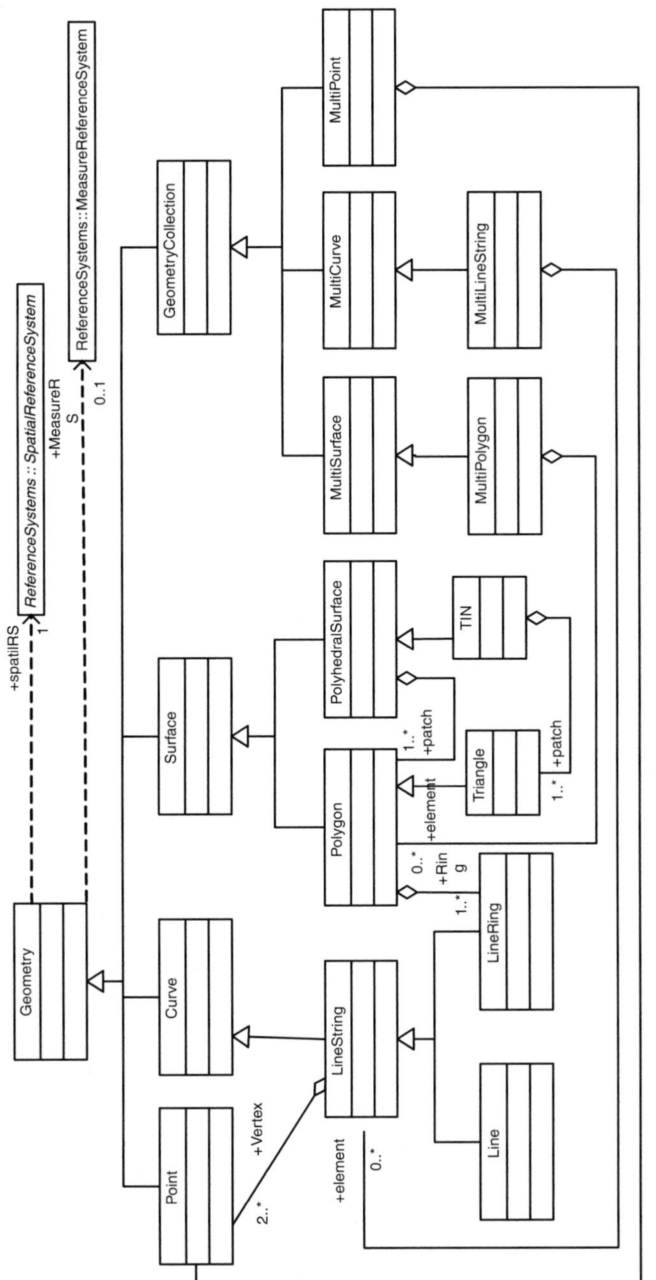

Fig. 6.1 SQL geometry type hierarchy (OGC 2006a)

and Geometry Collection. A Geometry Collection is a Geometry that is a collection of possibly heterogeneous geometric objects. MultiPoint, MultiCurve and MultiSurface are specific subtypes of a Geometry Collection used to manage homogenous collections of Points, Curves and Surfaces.

In the OGC's SFSQL Version 1.2.0, SQL functions are defined to construct instances of the above Types given well known text (WKT) or binary representations of the types, and the SQL/MM geometry methods are referenced to define the SQL functions on the above types.

6.4.1.2 SQL Implementation Using Pre-defined Data Types

In this implementation, a geometry-valued column is implemented using a "geometry ID" reference into a geometry table. A geometry value is stored using one or more rows in a single geometry table, all of which have the geometry ID as part of their primary key. The geometry table may be implemented using standard SQL numeric types or as SQL binary types. SDE belongs to this type.

The standard defines a schema for the management of feature tables, Geometry, and spatial reference system information in an SQL-implementation based on predefined data types. The corresponding tables are defined:

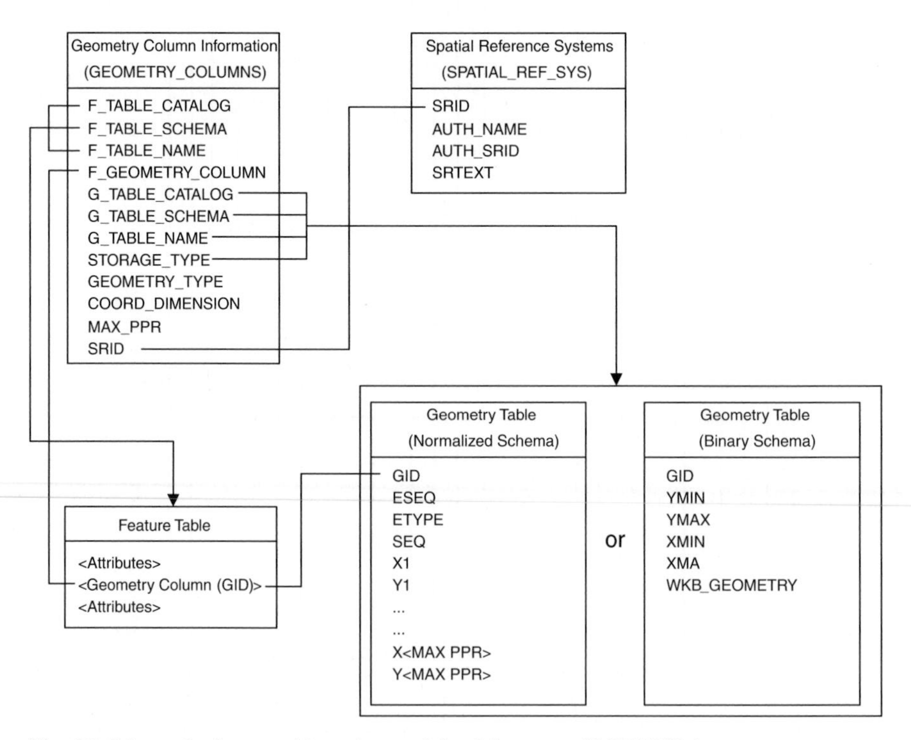

Fig. 6.2 Schema for feature tables using predefined data types (OGC 2006a)

1. The GEOMETRY_COLUMNS table describes the available feature tables and their geometry properties.
2. The SPATIAL_REF_SYS table describes the coordinate system and transformations for geometry.
3. The FEATURE TABLE stores a collection of features. A feature table's columns represent feature attributes, while rows represent individual features. The geometry of a feature is one of its feature attributes. While logically a geometric data type, a geometry column is implemented as a foreign key to a geometry table.
4. The GEOMETRY TABLE stores geometric objects, and may be implemented using either standard SQL numeric types or SQL binary types.

Depending on the storage type specified by the GEOMETRY_COLUMNS table, a geometric object is stored either as an array of coordinate values or as a single binary value. In the former case, predefined SQL numeric types are used for the coordinates and these numeric values are obtained from the geometry table until the geometric object has been fully reconstructed. In the latter case, the complete geometric object is obtained in the Well-known Binary (WKB) representation as a single value.

6.4.1.3 SQL Implementation Using Geometry Types

This comprises an SQL implementation that has been extended with a set of "Geometry Types", as mentioned above. In this environment, a geometry-valued column is implemented as a column whose SQL type is drawn from the set of Geometry Types. The mechanism for extending the type system of an SQL implementation is through the definition of User Defined Types (UDT).

This standard defines a schema for the management of feature tables, geometry, and a spatial reference system information in an SQL-implementation with a Geometry Type extension (Fig. 6.3). The corresponding tables are:

1. The GEOMETRY_COLUMNS table describes the available feature tables and their geometry properties.
2. The SPATIAL_REF_SYS table describes the coordinate system and transformations for geometry.
3. The FEATURE TABLE stores a collection of features. A feature table's columns represent feature attributes, while rows represent individual features. The geometry of a feature is one of the feature attributes, and is an SQL Geometry Type.

6.4.2 ISO SQL/MM Part 3: Spatial

The ISO SQL/MM Part 3: Spatial is the international standard that defines how to store, retrieve and process spatial data using SQL. The standard describes the geometry types and methods provided for each type, and defines the Information Schema, based on a Definition Schema. It also explains the underlying spatial concepts,

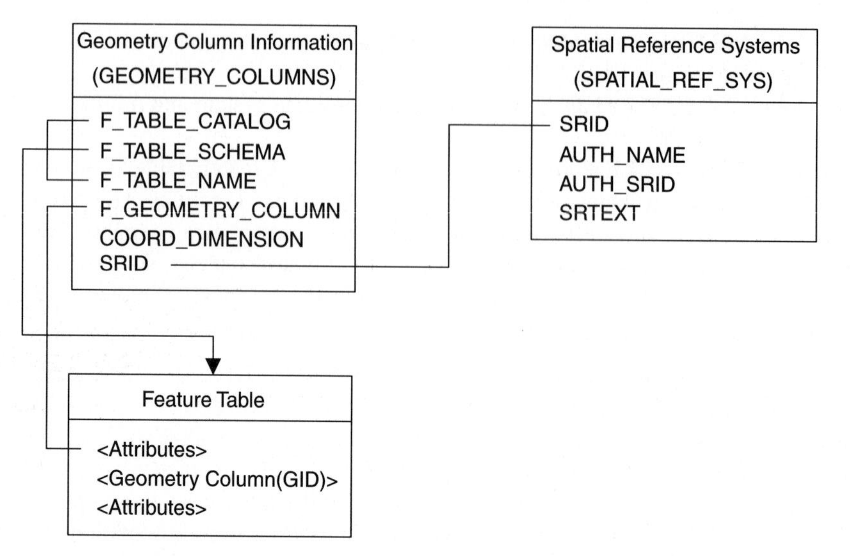

Fig. 6.3 Schema for feature tables using SQL with geometry types (OGC 2006b)

angles and direction handling, and states the coding and conformance rules for products that implement the standard.

6.4.2.1 SQL/MM Geometry Model

The OGC geometry class hierarchy is adapted for the corresponding SQL type hierarchy that is defined in the SQL/MM standard. Figure 6.4 shows the standardized

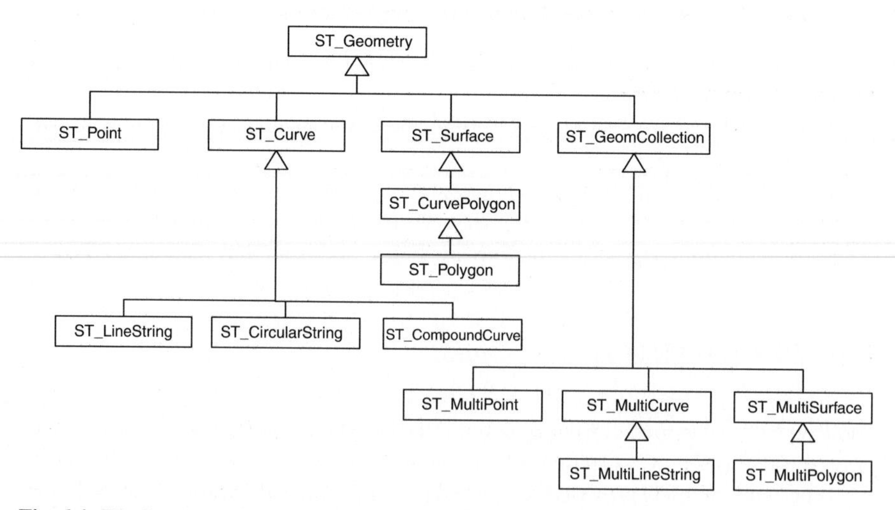

Fig. 6.4 ST_ Geometry type hierarchy diagram (ISO/IEC 2005)

geometry type hierarchy including subtype relationships between them. All types are used to represent geometric features in the 2-dimensional space (R2). Any geometry type can be used as the type for a column. Declaring a column to be of a particular type implies that any value of the type or of any of its subtypes can be stored in the column (ISO/IEC 2005).

The major differences between the SQL type hierarchy and the OGC geometry class hierarchy are the omission of the derived types Line and LinearRing (use values of type ST_LineString, which covers both cases), and the addition of a series of types (support circular arcs for example).

6.4.2.2 Geometry Methods

The majority of all geometry methods can be grouped into one of the following four categories (Stolze 2003):

1. Convert between geometries and external data formats.

   Three external data formats that can be used to represent geometries in an implementation-independent fashion include WKT, WKB, and geography markup language (GML). For example, ST_WKTToSQL returns the ST_Geometry value for a given WKT representation, and ST_AsText is provided for the conversion of a ST_Geometry value to WKT.

2. Retrieve properties or measures from a geometry. For example:

   ST_Dimension: returns the dimension of an ST_Geometry value.
   ST_GeometryType: returns the type of the ST_Geometry value.
   ST_SRID: observes and mutates the spatial reference system identifier.
   ST_IsMeasured: tests whether an ST_Geometry value has measures.

3. Compare two geometries with respect to their spatial relation. For example:

   ST_Equals: tests the spatial equality of two geometries.
   ST_Disjoint: tests whether two geometries do not intersect.
   ST_Intersects, ST_Crosses, and ST_Overlaps: tests whether the interiors of the geometries intersect.
   ST_Touches: tests whether two geometries touch at their boundaries, but do not intersect in their interiors.

   The methods return an INTEGER value, which is 1 if the spatial relation does exist, and 0 otherwise.

4. Generate new geometries from others. For example:

   ST_Buffer: generates a buffer at a speci_c distance around the given geometry.
   ST_ConvexHull: computes the convex hull for a geometry.
   ST_Difference, ST Intersection, and ST Union: constructs the difference, intersection, or union between the point sets defined by two geometries.

All methods and functions specified above operate in a 2D coordinate dimension space unless explicitly stated otherwise.

6.5 Typical Spatial Extensions of OSDBs

6.5.1 PostgreSQL-Based PostGIS

PostgreSQL belongs to the object-relational type of databases discussed above, and it is one of the most flexible databases available. It can be used for many purposes and supports a variety of programming languages to access and extend its core features. As an open object-oriented system, it can also be extended for a variety of data types, functions, operators, aggregates, index methods, and procedural languages.

PostGIS adds support for geographic objects to PostgreSQL. In effect, PostGIS "spatially enables" the PostgreSQL server, allowing it to be used as a backend spatial database for geographic information systems (GIS).

The strength of PostGIS is that it has become the standard backend for most FOSS4G tools (Ramsey 2006). As a result, a layer in PostGIS can be analyzed with GRASS, published over the web with MapServer, visualized on the desktop with OpenEV, and exported to proprietary formats with OGR. PostGIS is also used heavily by applications and libraries in the Java development language, via the standard Java Database Connectivity (JDBC) libraries.

6.5.1.1 Extending Methods

PostgreSQL provides an easy-to-use interface for users to add their own data types, functions, operators and so on (Ewald and Hans-Jürgen 2001). New spatial data types, functions, operators and aggregates can be created with SQL statements, and then implemented by lower level languages such as C. The PostgreSQL server can then incorporate user-written C code (compiled into shared libraries) into itself through dynamic loading. These procedures are described as follows:

1. *Adding a Data Type*: For many applications, self-defined data types are much more efficient than the DBMS's native data types. Extended data types (also known as User-Defined Types, UDT) not only speed up database applications significantly, but also allow the user to define functions that are not included in the DBMS's core distribution. As noted above, the most efficient way to add a new data type is to use the C programming language, because C is one of the fastest and most efficient programming languages available.

The most important components of a UDT are the in and out functions. The in function is used to insert values into the database while the out function is called when a value using the data type has to be displayed.

For spatial data management, the geometry data type is defined as follows:

```
CREATE TYPE geometry (
   internallength = variable,
   input = geometry_in,
   output = geometry_out,
   delimiter = ':',
   storage = main
);
Where the in and out functions are created as below:
CREATE FUNCTION geometry_in (GEOMETRY_OUT_REP)
   RETURNS geometry
   AS '@MODULE_FILENAME@', 'GEOMETRY_in'
   LANGUAGE 'C';
CREATE FUNCTION geometry_out (GEOMETRY_IN_REP)
   RETURNS cstring
   AS '@MODULE_FILENAME@', 'GEOMETRY_out'
   LANGUAGE 'C';
```

The GEOMETRY_in and GEOMETRY_out functions are C-defined in the shared library referenced by MODULE_FILENAME. In the C source file, the new-style (Version-1 Calling Convention) functions are defined as:

```
PG_FUNCTION_INFO_V1 (GEOMETRY_in);
Datum GEOMETRY_in (PG_FUNCTION_ARGS)
{
   Implementation......
}
and
PG_FUNCTION_INFO_V1 (GEOMETRY_out);
Datum GEOMETRY_out (PG_FUNCTION_ARGS).
{
   Implementation......
}
```

2. *Adding Functions*: Normally, user-defined functions (UDF) are added to PostgreSQL by using a loadable object (i.e., the shared library). Shared libraries are loaded at runtime (when the function is called the first time) and stay in memory for the rest of the session. Additional functions are not inserted and compiled by PostgreSQL automatically. Users add a function to a database by using the CREATE FUNCTION command.

Some geometry operation functions can be added as follows:

```
CREATE FUNCTION geometry_lt (geometry, geometry)
   RETURNS bool
   AS '@MODULE_FILENAME@', 'GEOMETRY_lt'
   LANGUAGE 'C';
CREATE FUNCTION geometry_gt (geometry, geometry)
```

```
    RETURNS bool
    AS '@MODULE_FILENAME@', 'GEOMETRY_gt'
    LANGUAGE 'C';
  CREATE FUNCTION geometry_eq (geometry, geometry)
    RETURNS bool
    AS '@MODULE_FILENAME@', 'GEOMETRY_eq'
    LANGUAGE 'C';
  CREATE FUNCTION geometry_cmp (geometry, geometry)
    RETURNS integer
    AS '@MODULE_FILENAME@', 'GEOMETRY_cmp'
    LANGUAGE 'C';
```

Where the GEOMETRY_lt (less than), GEOMETRY_gt (great than), GEOMETRY_eq (equal) and GEOMETRY_cmp (compare) functions are also C-defined.

3. *Adding Operators*: To make the geometry data type more powerful, additional functions have to be implemented and assigned to operators. To take the CREATE OPERATOR command as an example, an operator can be added to a database. Some sorting operators (<, > and =) for the geometry type can be defined as:

```
  CREATE OPERATOR < (
    LEFTARG = geometry, RIGHTARG = geometry,
  PROCEDURE = geometry_lt,
    COMMUTATOR = '>', NEGATOR = '>=',
    RESTRICT = contsel, JOIN = contjoinsel
  );
  CREATE OPERATOR > (
    LEFTARG = geometry, RIGHTARG = geometry,
  PROCEDURE = geometry_gt,
    COMMUTATOR = '<', NEGATOR = '<=',
    RESTRICT = contsel, JOIN = contjoinsel
  );
  CREATE OPERATOR = (
  LEFTARG = geometry, RIGHTARG = geometry,
  PROCEDURE = geometry_eq,
    COMMUTATOR = '=',
    RESTRICT = contsel, JOIN = contjoinsel
  );
```

Every operator is "syntactic sugar" for a call to an underlying function (specified by the PROCEDURE clause) that does the real work. The underlying function must first be created before the operator is called. The procedure clause and the argument clauses are the only required items in CREATE OPERATOR. The COMMUTATOR, RESTRICT and JOIN clause shown in the example are optional hints to the query optimizer.

4. *Adding Aggregates*: Aggregate functions in PostgreSQL are expressed as state values and state transition functions. That is, an aggregate can be defined in terms

of a state to be modified whenever an input item is processed. To define a new aggregate function, a data type is selected for the state value, an initial value for the state, and a state transition function. The state transition function is just an ordinary function that could also be used outside the context of the aggregate. A final function can also be specified, in case the desired result of the aggregate is different from the data that need to be kept in the running state value.

For example, the extent aggregate for all geometries in a spatial table is defined as:

```
CREATE AGGREGATE extent (
    sfunc = combine_bbox,
    basetype = geometry,
    type = box2d
);
```

In summary, PostgreSQL uses an "SQL definition + C implementation" extending method to create new data types, functions, operators and aggregates, in which C symbols are linked with SQL definitions by an AS 'filename' clause in SQL statements, such as CREATE FUNCTION.

6.5.1.2 Using PostGIS – Examples

With the PostGIS extention, database users or developers can access and operate spatial data with extended SQL. Typically, a spatially-enabled table should be created first in two stages:

1. Create a normal non-spatial table.
   For example: CREATE TABLE roads_geom (ID int4, NAME varchar(25)).
2. Add a spatial column to the table using the OpenGIS "AddGeometryColumn" function.

The syntax is: AddGeometryColumn (<schema_name>, <table_name>, <column_name>, <srid>, <type>, <dimension>).

Note that the data type name of the geometry column is "geometry" in PostGIS and its storage structure is based on extended formats of OGC WKB. After spatial columns are created, GIS data can be bulk loaded by piping a large text file full of SQL "INSERT" statements into the SQL terminal monitor. A data upload file (gistab.sql for example) might look like this:

```
BEGIN;
INSERT INTO roads_geom (ID, GEOM, NAME) VALUES -
(1, GeomFromText ('LINESTRING (19 22,11 24)', - 1),'Jeff Rd');
INSERT INTO roads_geom (ID, GEOM, NAME) VALUES -
(2, GeomFromText('LINESTRING (14 28,16 27)', - 1),'Geo Rd');
COMMIT;
```

After populating spatial columns with values, a spatial index can be built for the data. PostgreSQL supports three kinds of indexes by default, namely B-Trees, R-Trees, and Generalized Search Trees (GiST). GiST indexes break up data into “things to one side”, “things which overlap”, ”things which are inside” and they can be used on a wide range of data-types, including GIS data. PostGIS uses an R-Tree index implemented on top of GiST to index GIS data. The spatial (GiST) indexes are built with statements like the following:

```
CREATE INDEX [indexname]
ON [tablename]
USING GIST ([geometrycolumn] GIST_GEOMETRY_OPS);
```

With the assistance of a spatial index, queries can be executed using SQL. Spatial operators can be used to do comparisons and queries on spatial tables. The most straightforward means of pulling data out of the database is to use a SQL select query:

```
SELECT * FROM roads_geom
WHERE GEOM && ‘BOX3D (11 24,19 43)’ :: box3d;
```

All OGC functions are implemented in PostGIS and can be used in SQL or PL/PGSQL. These include management functions such as AddGeometryColumn and DropGeometryColumn, geometry relationship functions such as Intersects and Crosses, geometry processing functions such as GeomUnion and Buffer, geometry accessors such as AsText and AsBinary, geometry constructors such as GeomFromText. For example, we can calculate total length of all roads, expressed in kilometers as follows:

```
SELECT Sum(Length(the_geom))/1000 AS km_roads
FROM roads_geom;
```

PostGIS layers can also be added to mapfiles used by MapServer, but only if MapServer is compiled to do so. To add PostGIS layers in the MapServer map file the syntax is as follows:

```
LAYER
  CONNECTIONTYPE postgis
  NAME “roads”
  CONNECTION “host=ip_ADDR user=dbuser dbname=gisdb”
  DATA “geom from roads”
  ...
END.
```

6.5.2 *MySQL Spatial Extension*

MySQL (Version 5.0) implements a subset of SQL with the Geometry Types environment proposed by the OGC. This differs from PostGIS, which has only one SQL type of "geometry" with EWKB format. Internally, MySQL stores geometry values in a format that is not identical to either the WKT or WKB format.

MySQL Spatial Extension describes a set of SQL geometry types, as well as functions on those types to create and analyze geometry values. MySQL 5.0 has limitations on functions that test spatial relationships between geometries, and some functions designed to implement spatial operators are not yet implemented. However, they may appear in future releases.

6.5.2.1 MySQL Spatial Data Types

MySQL has data types that correspond to OGC classes, as shown in Fig. 6.5. The data types start from the most generic at the top of the hierarchy, GEOMETRY, to a number of specific types, such as POINT and LINESTRING. Some of the data types are "abstract", such as the GEOMETRY type.

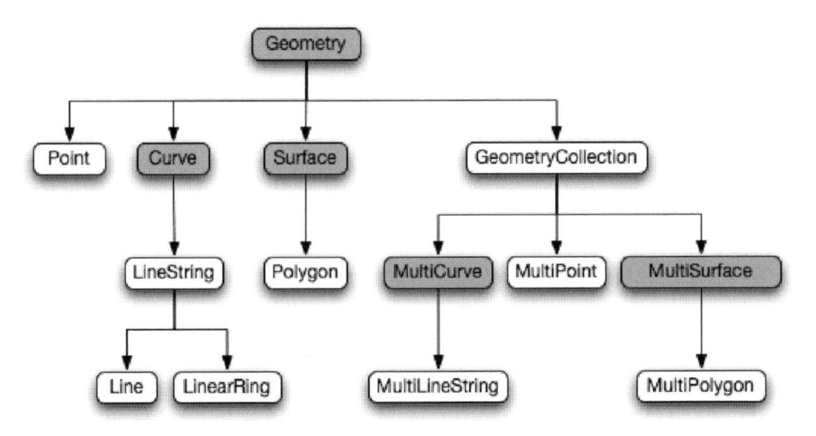

Fig. 6.5 MySQL spatial data types (abstract types in gray) (Karlsson 2004)

6.5.2.2 Using MySQL Spatial Extension – Examples

MySQL provides a standard way of creating spatial columns for geometry types. For example, a table with bus station name and location information can be created with the CREATE TABLE statement:

```
CREATE TABLE bus_station (
name CHAR(80) NOT NULL,
geom POINT NOT NULL,
PRIMARY KEY(name),
);
```

Next spatial columns are created and populated them with spatial data. Values should be stored in internal geometry format, but these can be converted from either WKT or WKB format. Similar to PostGIS, the easiest way to do this is to use the GeomFromText() function, that takes a string in WKT format and converts it to a spatial object.

```
INSERT INTO bus_station VALUES ('A street', Geom-
FromText('POINT(234.678 564.121)'));
INSERT INTO bus_station VALUES ('B street', Geom-
FromText('POINT(337.322 366.124)'));
......
```

After populating spatial columns with values, a spatial index can be built for the data. MySQL uses R-Trees with quadratic splitting to index spatial columns, which is one of the standard methods of building an R-tree index. The syntax use is similar to creating regular indexes, but extended with the SPATIAL keyword. The following examples demonstrate how to create spatial indexes:

```
CREATE SPATIAL INDEX sp_index ON bus_station (geom);
```

Also, using the CREATE TABLE statement directly:

```
CREATE TABLE bus_station (
name CHAR(80) NOT NULL,
geom POINT NOT NULL,
PRIMARY KEY(name),
SPATIAL INDEX(geom)
);
```

With the prepared spatial data, MySQL provides a set of functions and operators to perform queries and operations, but not all OGC functions are implemented yet. Functions that describe relations between two geometries (such as Contains(), Crosses()) are mainly based on MBR (Minimum Bounding Rectangle) of geometries. Spatial operators that create new geometries from existing ones (such as Buffer(), ConvexHull()) are not implemented in MySQL. They may appear in future releases.

6.5.3 Comparison of PostGIS, MySQL and Oracle Spatial

Based on the above discussion, Table 6.3(a)(b) provides a comprehensive comparison of the main features that characterise PostGIS, MySQL Spatial Extension and Oracle Spatial, in which Oracle Spatial is selected as the commercial DB representative.

Table 6.3 Comparison of spatial DBMSs (adapted from Lemmen 2007) (N/A = information not available)

	(a)		
	MySQL	PostGIS	Oracle
DBMS/ Date of Introduction	MySQL 5.0/2007.1	PostGIS 1.2.1/2007.1	Oracle 10g R2 /2004.1
Standardization			
OGC compliant	SQL with Geometry types. Compliant with the exception of precise spatial operations	SFSQL-TF 1.1 Certified	SFS1; GML 2.0; OLS 1.1; SRS; WMS 1.1
ISO standards support	N/A	SQL/MM	N/A
Spatial data types, operators and indexing			
Spatial data-types (vector-oriented) 2D, 3D	2D only. Rtree keys	As specified in OGC SFSQL: Point, Linestring, Polygon, Multipoint, Multilinestring, Multipolygon, Geomatrycollection	All types supported in SFS1 + circles, arcs, combinations of arcs and lines and rectangles; Support for 3D storing of lines/points/polygons
Spatial data-types (raster-oriented)	N	CHIP datatype to store rasters in PostgreSQL	Grid-based data and image data are supported using GeoRaster data type
Spatial data-operators (vector-oriented)	N	OGC + ST_* + PostGIS Specific – several hundred data-operators are supported	Inside, overlap, intersect, (Egenhofer) + Within Distance, Nearest Neighbour+ area, length + mbr, centroid, etc. + union, intersect, or, xor + aggregates
Spatial data-operators (raster-oriented)	N	Input/output only	Crop, Scale, Transform and more than 110 other functions
Spatial data index 2D, 3D	2D R-Trees with quadratic splitting	2D GiST and R-Tree (with multi-version concurrency and recovery)	R-Tree (supports 3D) and Quad-Tree

Table 6.3 (continued)

	(a)		
	MySQL	PostGIS	Oracle
Supported co-ordinate systems/projections	N	All EPSG CRS systems (this means most systems and projections are covered)	Forthcoming release will support 3D coordinate systems, a new geometry type for 3D data, a point-cloud data type and TIN data model

	(b)		
	MySQL	PostGIS	Oracle
Functionality			
Topology support (node, edge, face)	N	SQL/MM Topology Model + Basic Functions	Full support for planar topology
Linear networks support	N	N	N/A
Linear referencing	N	Y	Y, linear referencing can support 4D
Spatial join algorithms	Partially	Spatial join using index operations, spatial/ attribute query optimisation	N/A
Support of spatio/temporal models	N	User maintenance	Versioned tables with Workspace Manager
Data exchange			
Exchange formats	SHP	All FME formats	N/A
XML, GML, CityGML, X3D, KML output	N	GML, SVG, KML	GML 2.0 and much of GML 3.0. OGCs Open Location Services. Forthcoming: WFS-T, Web Catalog Service, OpenLS
Platform			
64 bit platform support	N/A	Y	Y
Platforms	Linux (many flavours), Windows, Mac OSX, Solaris, IBM AIX, HP-UX, FreeBSD, SCO OpenServer	AIX, BSD/OS x86, GNU/Linux, HP-UX, IRIX MIPS, Mac OS X, NetBSD x86 Solaris, Tru64 UNIX Alpha, UnixWare x86, Windows x86	All major Unix including Linux, HP-UX, Solaris, AIX, Windows, IBM Mainframes

6.6 Future Directions

This section concludes this chapter with a discussion of some of the gaps that currently remain in terms of spatially-enabled databases. In particular, the challenges for supporting topological relationships in vector databases, spatio-temporal data models, and raster data formats are outlined.

6.6.1 Making the Best of the Object Model and Topological Model

Spatial characteristics of a feature can be described by an object model (view) or topological model separately or in combination. In the object model, spatial characteristics are usually described by geometric objects with coordinates using simple geometry storage. In this context, spatial representation is object-oriented, e.g., a river crossing two regions will not break into two primitives for storage.

According to Clementini et al. (1993), topological relationships are the subset of spatial relationships characterized by the property of being preserved under topological transformations, such as translation, rotation and scaling. Hence, the topological model represents spatial objects (point, line, and area features) using an underlying graph of topological primitives, such as nodes, edges and faces. These primitives, together with their relationships to one another and to the features whose boundaries they represent, are defined by representing the feature geometries in a planar graph of topological elements.

This means that there are two potential methods used when working with features, namely one in which features are defined by their coordinates and another in which features are represented as an ordered graph of their topological elements. The topology model is mainly employed to define and enforce data integrity (e.g., no gaps should exist between parcel features, parcels should not overlap) and support topological relationship queries and navigation, but its data structure is complex and features have to be assembled on the fly when they need to be queried or analyzed. On the other hand, the data structure (physical schema) of the object model with simple geometry storage is simple, fast, and scalable, and it is easy for programmers to write interfaces as well. Its disadvantage is that maintaining the data integrity readily provided by topology is not as easy to implement for simple geometry (ESRI 2005). As a consequence, users have applied one data model for editing and maintenance and used another for deployment and management.

According to OSDB development, the first step in building a spatial DBMS is by having data types and operators for simple features (i.e. geometric primitives) including point, line and polygon with an object view. This has reached a level of standardisation and is now implemented in a lot of open source DBMSs, including PostgreSQL, Ingres and MySQL. The next step is to have support for topologically structured features in DBMS. Due to unresolved issues as to how to implement effectively topological models within a relational or object-relational DBMS (Oosterom et al. 2002), there are no standard implementation specifications for the topology model. Hence, few OSDBs have full topology support. For example, PostGIS (V1.1.6) has preliminary support for topology and TerraLib's physical model (V3.1.4) supports node-arc topology.

Except for the explicit storage of topology primitives (i.e., the nodes, faces and edges) and their relationships, there is topology implementation in the logical level, such as the rule-based topology model introduced by ESRI's ArcGIS Geodatabase model. It uses simple geometry storage for all features in the Geodatabase and to index the shared coordinates on each feature using a specialized topological index (ESRI 2005). It retrieves the topological primitives on the fly when they are needed along with their relationships. The comparison on rule-based and explicit storage of the topology structure along with their advantages and disadvantages are shown in (Baars 2003).

In effect, making the best of the object model and the topological model has been considered as more than a data storage problem. It needs to be considered as part of a complete spatial database solution. This solution includes a complete data model for objects, operators, rules and tools, and the ability to work with features (points, lines, and polygons) as well as topological elements (nodes, edges, and faces) and their relationships to one another. Other significant issues, such as performance, scalability, transaction isolation, multi-user editing, distributed processing, should also to be considered in topology model design. Failure of an organization to consider these issues would be a critical mistake.

6.6.2 Spatio-Temporal Extensions

Many geographic features in the real world have attributes describing location and time. For example, global change (as in climate and land use change), transportation systems (traffic surveillance data, intelligent transportation systems, vehicle tracking), social change (demographic composition of populations, health, etc.), telecommunications (mobile phones), and multimedia applications (e.g., animated movies).

Spatio-temporal databases deal with applications where data types are characterized by both spatial and temporal semantics (Pelekis et al. 2004). Development and research in this area started decades ago, when management and manipulation of data relating to both spatial and temporal changes were recognized as indispensable. However, spatio-temporal data handling is not a straightforward task due to the complexity of the data structures, together with the representation and manipulation of the data involved. Earlier work in this area began from separate research and development threads in both temporal and spatial databases.

To deal with spatio-temporal data, one alternative is to build a specialized DBMS created for efficient support of spatio-temporal data types, as in the CONCERT (Relly et al. 1997) and SECONDO (Dieker and Güting 2000) projects. When it is not possible to use a specialized DBMS, another way is to extend commercial or OS object-oriented or object-relational DBMS for storage and manipulation of complex objects of spatio-temporal data. Traditional relational database technology is not suitable for managing spatio-temporal data, which are multi-dimensional with complex structures and behaviours (Wang et al. 2000). Queries such as finding land parcels which are within certain areas with certain land use types during a given time period cannot be efficiently answered by pure relational technologies. Such queries involving spatial and temporal relationships need to be supported by new spatial data models, new object storage structures, multi-dimensional indexing mechanisms, and query evaluation strategies based on new types of database architecture. Though object-oriented and object-relational DBMSs have extensible capacity and mechanism to storage and manipulation of complex objects such as spatio-temporal data (Wang et al. 2000), no applicable OS DBMS with a spatio-temporal extension yet exists.

Another alternative is to build a layered architecture on top of an existing object-relational DBMS. TerraLib's (see Chap. 12) spatio-temporal database uses this approach. TerraLib includes different spatio-temporal data types (event, mutable objects, moving objects) and supports spatio-temporal queries. The descriptive, spatial and temporal components of geographical objects are stored separately on the database and are linked by the object ids. Table 6.4 is a comparison of TerraLib's spatio-temporal support with a commercial library (ArcSDE) and another open source one (PostGIS).

Since there is no international standard and widely accepted spatio-temporal data model with well-developed theories and technologies, it is still a long way to go to develop a general spatio-temporal DBMS. Up to now, a main solution of spatio-temporal applications is to manage the data in a spatial DBMS and then interpret the domain related spatio-temporal semantics in the application level.

6.6.3 Raster Data Support

The emergence of high-resolution remote sensors and the establishment of digital photogrammetry have lead to a rapid increase in the amount of spatial data being acquired as raster images. In addition to this large volume of raster imagery, a number of other spatial raster types, such as raster maps, regular matrix digital terrain models and thematic rasters or "grids" have become an important information

Table 6.4 Spatio-temporal support comparison (Camara et al. 2004)

Requirement	ArcSDE	PostGIS	TerraLib
DBMS supported	Oralce, DB2, SQL Server, Informix	PostgreSQL	Oracle, PostgreSQL, MySQL
Spatio-temporal support	No	No	Yes
Dynamic modeling	No	No	Yes
Iterates over spatio-temporal data structures	No	No	Yes

source for GIS. Thus, it is increasingly important to build databases capable of dealing with these raster types together with other spatial and non-spatial data types.

Most spatial databases can deal with vector geometries (e.g., polygons, lines and points), but have limited facilities for handling raster data. To build spatial databases that handle raster data types, one alternative is to develop specialized data servers, as in the case of PARADISE (Patel et al. 1997) and RASDAMAN (Reiner et al. 2002). The chief advantage of this approach is the capacity of performance improvements, especially in the case of large image databases. The main drawback of this approach is the need for a specialized, non-standard server, which would greatly increase the management needs for most GIS applications (Vinhas et al. 2003). Other ways are using standard DBMS, such as a middleware-based spatial raster management architecture and spatial raster management architecture based on an ORDBMS server extension. By means of adequate indexing, compression and retrieval techniques, satisfactory performance can be achieved using a standard DBMS, even for very large raster images. Thus, using a standard DBMS with a spatial extension has become the main trend to manage massive raster data.

Remote sensing images have a large variety of spatial and spectral resolutions, and due to the update frequency, raster databases also tend to be very large. As an example of the rate of accumulation, consider the IKONOS platform from Space Imaging. Once every 98 min (14 times a day), Space Imaging's IKONOS satellite circles the globe collecting images of the Earth. This nature of spatial raster data requires a number of aspects to be considered when designing a spatial raster DBMS which should support this data type. These include spatial partitioning and partition indexing based on compromises between partition size and performance, different raster structures for building multi-resolution pyramids, the organization of spatial raster attributes, and raster data storage with different compression techniques, such as wavelets. Moreover, the scenario of very large raster mosaic management represents a typical example for current and future data management requirements, both in terms of data volume and functionality. Raster mosaics consist of lateral combinations of multiple raster objects into "seamless" raster databases. In the past, raster mosaics were limited in size to a few GB, mainly due to the unavailability of suitable data management solutions.

Several commercial spatial databases support raster data management, such as Oracle Spatial 10g with GeoRaster and ArcGIS Geodatabase. However, with the exception of Terralib (see Chap. 12) the FOSS4G world lags somewhat behind on this aspect. Hence, it is clear that raster support is one of the future projects for OSDBs.

References

Ashworth M, Cotton P, Kucera H, Brien D (1998) Comparison of spatial schema in ISO/TC 211, OGC, and SQL/MM, ISO Expert Contribution

Baars M (2003) A comparison between ESRI geodatabase topology and laser-scan radius topology. Available: http://repository.tudelft.nl/file/391797/371192

Borrie H (2006) Firebird databases as the back-end to enterprise software systems. Firebird Enterprise White Paper. Available: http://www.firebirdsql.org/devel/doc/papers/html/paper-fbent.html

Breunig M, Türker C, Böhlen MH, Dieker S, Güting RH, Jensen CS, Relly L, Rigaux P, Schek H-J, Scholl M (2003) Spatio-temporal databases: The CHOROCHRONOS approach – Chapter 7. Springer-Verlag, Heidelberg

Camara G, CartaxoR, Monteiro A, Ferreira K, Vinhas L, Carvalho M, Casanova M (2004) TerraLib: An open source GIS library for spatio-temporal databases. Available: http://www.dpi.inpe.br/gilberto/papers/terralib_may2004.pdf, Unpublished manuscript

Clementini E, Felice PD, Oosterom P (1993) A small set of formal topological relationships suitable for end-user interaction. In: Abel D, Ooi BC (eds), Advances in spatial databases -Third International Symposium, SSD'93, Singapore. Springer, Berlin, pp 277–295

Dieker S, Güting RH (2000) Plug and play with query algebras: SECONDO, a Generic DBMS development environment. In: Proc. of the Int. Database Engineering and Applications Symp., Yokohama, Japan, pp 380–392

ESRI White Paper (2005) GIS topology. Available: www.esri.com/library/whitepapers/pdfs/gis_topology.pdf

Ewald G, Hans-Jürgen S (2001) PostgreSQL developer's handbook. Sams Publishing, USA

Fabalabs Research Paper (2005) Comparison of the enterprise functionalities of open source database management systems. Available: http://www.fabalabs.org/research/papers/FabalabsResearchPaper-OSDBMS-Eval.pdf

Ingres (2006) Ingres 2006 release summary. Available:http://downloads.ingres.com/download/rs.pdf

ISO/IEC (2005) Information technology—SQL multimedia and application packages – part 3: spatial 3rd ed

Karlsson A (2004) GIS and spatial extensions with MySQL. Available: http://dev.mysql.com/tech-resources/articles/4.1/gis-with-mysql.html

Lemmen C (2007) Product survey on geo-databases. GIM International, 21(5)

MaxDB (2005) MaxDB – The professional DBMS (MaxDB Whitepaper). Available: http://dev.mysql.com/doc/maxdb/pdf/whitepaper.pdf

OGC (2006a) OpenGIS implementation specification for geographic information – simple feature access – Part 1: Common Architecture

OGC (2006b) OpenGIS implementation specification for geographic information – simple feature access – Part 2: SQL Option

Oosterom P, Stoter JE, Quak CW, Zlatanova S (2002) The balance between geometry and topology. In: Richardson D, Oosterom P (eds) Advances in spatial data handling, 10th International Symposium on Spatial Data Handling, Springer-Verlag, Berlin, pp 209–224

Patel JM, Yu JB, Kabra N, Tufte K, Nag B, Burger J, Hall NE, Ramasamy K, Lueder R, Ellman C, Kupsch J, Guo S, DeWitt DJ, Naughton J (1997) Building a scalable geo-spatial DBMS: technology, implementation, and evaluation. Proceedings ACM SIGMOD Conference, 336–347.

Pelekis N, Theodoulidis B, Kopanakis I, Theodoridis Y (2004) Literature review of spatio-temporal database models. Knowledge Eng Rev J, 19(3):235–274

Ramsey P (2006) The state of open source GIS. Refraction Research. Available: http://www.refractions.net/white_papers/oss_briefing/2006-06-OSSBriefing.pdf

Reiner B, Hahn K, Höfling G, Baumann P (2002) Hierarchical storage support and management for large-scale multidimensional array database management systems. In: 3th International Conference on Database and Expert Systems Applications (DEXA). Aix en Provence, France

Relly L, Schek H-J, Henricsson O, Nebiker S (1997) Physical database design for raster images in CONCERT. In: 5th International Symposium on Spatial Databases (SSD'97). Springer, Berlin, pp 259–279

Stolze K (2003) Sql/mm spatial – the standard to manage spatial data in a relational database system. In: Weikum G, Schöning H, Rahm E (eds) BTW 2003. pp 247–264

Vinhas L, Souza RCM, Camara G (2003) Image data handling in spatial databases. In: Geoinfo 2003. INPE

Wang X, Zhou XF, Lu SL (2000) Spatiotemporal data modeling and management: a survey. In: Proceedings of the 36th International Conference on Technology of Object-Oriented Languages and Systems (TOOLS-Asia'00). IEEE Press, pp 202–211

Chapter 7
MapGuide Open Source

Robert Bray

Abstract MapGuide Open Source is a modern Web-based geospatial platform. This chapter provides an introduction to the MapGuide Open Source project, the MapGuide software components, a brief introduction to creating applications with MapGuide, and concludes with a brief discussion of what the future of this project may hold.

7.1 Introduction

MapGuide Open Source is a server-based geospatial platform that enables users quickly to develop and deploy Web mapping applications and spatial data centred Web services. It enables the creation of very rich Web applications that offer functionality that is typically only available in desktop geographic information system (GIS) software. Throughout the remainder of this chapter MapGuide Open Source software is referred to simply as MapGuide.

MapGuide features an interactive viewer that includes support for feature selection, property inspection, map tips, and operations such as buffer, select within, and measure. On the server end MapGuide includes an XML database for storing and managing content, and supports most popular spatial file formats, databases, and standards. The MapGuide platform can be deployed on Linux or Microsoft Windows, supports Apache and IIS Web servers, and offers extensive PHP, .NET, Java, and JavaScript application programming interfaces (APIs) for developing applications.

Robert Bray
Autodesk, Inc., 2100, 645 - 7th Ave SW, Calgary, Alberta T2P 4G8 Canada
e-mail: robert.bray@autodesk.com

G.B. Hall, M.G. Leahy (eds.), *Open Source Approaches in Spatial Data Handling*.

7.1.1 Project Mission and Goals

The mission of the MapGuide Open Source community is to create the leading international Web-based platform for developing and deploying Web mapping applications and spatial data-centred Web services. The goals set out for the platform are as follows:

- Use a service-oriented architecture pattern.
- Make the platform fast, scalable, and secure.
- Make use of existing open source (OS) components wherever possible.
- Support rich read/write access to both vector and raster spatial data.
- Provide a full suite of spatial analysis capabilities.
- Produce visually stunning cartographic maps.
- Include viewers that work with any browser on any platform.
- Provide the highest degree of map interactivity with the thinnest client possible.
- Conform to open spatial data standards.

7.1.2 Project History

MapGuide was first introduced in 1995 by Argus Technologies of Calgary, Alberta as a proprietary product called Argus MapGuide. Autodesk, Inc. acquired Argus Technologies in the fall of 1996 and within a few months the first release under the Autodesk brand was introduced as Autodesk MapGuide 2.0. The software progressed through a number of releases leading up to the currently available Autodesk MapGuide version 6.5. To this day MapGuide 6.5 and previous releases are known for their ease of deployment, rapid application development, data connectivity, scalability, and overall performance.

Despite its success the MapGuide 6.5 architecture has some inherent limitations. For example, most MapGuide applications depend upon a client-side Plug-in, ActiveX Control, or Java Applet with much of the application logic written in JavaScript using APIs exposed by the client-side plug-in. All spatial analysis is performed client-side on rendered graphics rather than on the underlying spatial data. Finally, the server platform is only available on Microsoft Windows.

In the spring of 2004 a dedicated team of developers began work on what is now MapGuide Open Source. The goals of the team were simple, namely to retain all of the best aspects of MapGuide 6.5, while also meeting the goals set out in the previous section. Autodesk released this version of the software under the Lesser General Public License (LGPL) in November 2005, and contributed the code to the Open Source Geospatial Foundation (OSGeo) in March 2006. During the remainder of 2006 the MapGuide Open Source project formed a project steering committee and adopted an open collaborative development model. In February of 2007 the project graduated from the OSGeo incubation process and became a fully endorsed project of OSGeo. Since its release, the MapGuide community has been steadily growing, attracting a wide range of users and application developers, and is now seeing active community-driven contributions and enhancements.

7.2 MapGuide Component Architecture

MapGuide is a multi-tier Web-based GIS capable of supporting multiple types of clients via standard internet protocols. The components of MapGuide are shown in Fig. 7.1.

The server tier consists of one or more MapGuide Servers that perform all the heavy lifting such as data query and update, map rendering, and more. Additional servers can be added as the number of concurrent users of a site grows. As shown, MapGuide Servers are typically networked with file and database servers that store spatial and attribute data.

The Web tier consists of one or more Web servers working in conjunction with the MapGuide Web Server Extensions, which intercept requests from clients and process them with the assistance of a MapGuide Server. The MapGuide Web Server

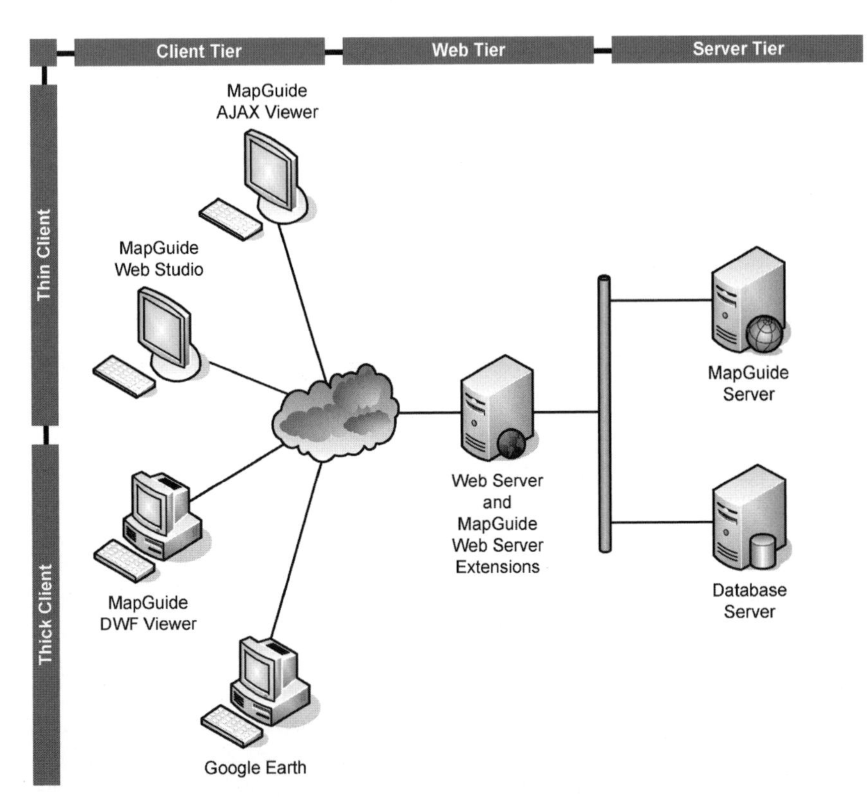

Fig. 7.1 MapGuide component architecture

Extensions expose a rich API that can be utilized from PHP, ASP.NET, or Java-based Web applications running within the Web server environment.

The client tier supports both thick and thin clients. MapGuide includes support for two viewers that run within a Web browser, one developed with Asynchronous JavaScript and XML (AJAX) technology and the other developed utilizing the freely available Autodesk Design Web Format (DWF) Viewer. For defining data sources, layers, and maps MapGuide includes a Web-based authoring application called Web Studio. In addition, Google Earth and applications that support the Open Geospatial Consortium Web Map Service (OGC WMS) or Web Feature Service (WFS) standards can be MapGuide clients.

Note that for application development, demonstrations, and deploying field/mobile applications, all of the MapGuide components can be installed and configured on a single laptop or workstation.

7.3 MapGuide Server Tier

The MapGuide server tier consists of one or more MapGuide Server instances which are typically networked with file and database servers that store spatial and attribute data. The MapGuide Server software runs as a daemon on various distributions of Linux or as a Service on Microsoft Windows. Having the MapGuide Server run as a separate long-lived process from the Web server has the following advantages:

- A long running process can intelligently cache information in memory and pool expensive resources such as database connections.
- System administrators can tune site performance and scalability by independently allocating physical hardware to either the Web tier or the server tier as required. In typical deployments more hardware will need to be allocated in the server tier for map rendering than in the Web-tier.
- A firewall can be placed between the Web tier and the server tier, ensuring that any sensitive data remain behind the firewall and thus are not co-located with publicly accessible Web servers.

The structure of the MapGuide Server itself is shown in Fig. 7.2. The Core Server Framework consists of a collection of components that facilitate the handling of requests, resource pooling, caching, and other functions common to all server software. These core framework components are based upon the Adaptive Communications Environment (ACE) which is an OS library that implements many core patterns necessary for developing highly scalable concurrent communication software (http://www.cs.wustl.edu/~schmidt/ACE.html).

Of greater interest is the collection of Services shown in the bottom half of Fig. 7.2 that exposes the spatial functionality of the MapGuide Server. These services support the definition of spatial data content, data query and update, and the rendering of maps in various formats. The next section describes each of these services in more detail.

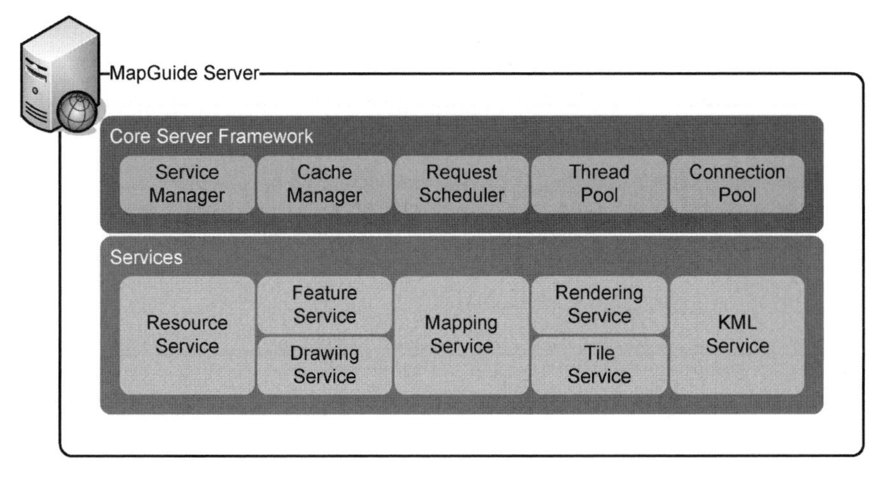

Fig. 7.2 MapGuide server components

7.4 MapGuide Services

The MapGuide Server exposes its functionality as a collection of services. Each service is defined as a logical set of operations that appear as APIs in the Web Tier, giving the application developer full access to the functionality of the MapGuide Server. The MapGuide Services and the operations they expose follow the definition of a service as defined in a Service Oriented Architecture (SOA).

7.4.1 Resource Service – Geospatial Content Management

The Resource Service provides operations for storing, retrieving, and managing spatial data in the form of XML based documents that MapGuide calls "Resources". These XML documents provide the instructions that tell other services to how to connect to data, how to render data as a layer, and how to combine layers into maps. Oracle's OS Berkeley DB XML is used as an embedded database to store and manage the XML resource documents, in a familiar directory-like paradigm for organizing content (http://www.oracle.com/database/berkeley-db/xml/index.html).

Table 7.1 lists each of the key resource document types, and describes what they define and how they relate to one another. Storing these resource documents inside an XML database allows MapGuide to understand and track the dependencies between resource documents. The referencing model also allows resources to be shared, e.g. a layer definition can be referenced by several map definitions and updates to the layer definition are immediately reflected in all of the maps that reference the layer. References to a resource document are specified using a Uniform Resource Locator (URI) type of construct as follows: Library://MyLayers/

Table 7.1 Key resource document types

Resource type	Description
FeatureSourceDefinition	Defines a connection to a source of feature data, (e.g., a directory of ESRI shape files or a connection to a PostGIS database).
DrawingSourceDefinition	Defines a connection to an Autodesk DWF file, which is used to incorporate computer assisted design (CAD) data into a map.
LayerDefinition	Defines how to style a particular feature class at one or more scale ranges. It contains a reference to a feature or drawing source definition, a list of one or more scale ranges, and rules for stylizing and labelling features at each scale. The Style rule may embed or refer to one or more simple or compound symbol definitions.
SimpleSymbolDefinition	Defines a symbol for stylizing feature data in terms of either an existing PNG format image or vector graphics defined within the XML. Attributes of the symbol such as color, or thickness can be parameterized and hence defined on a per feature basis.
CompoundSymbolDefinition	Defines a compound symbol made up of one or more simple symbol definitions.
MapDefinition	Defines a map in terms of a coordinate system, extent, background color, and a hierarchical grouping of layers. Each layer is defined as a reference to a layer definition resource.
WebLayout	Defines the layout of the map and user interface elements within the client browser. User interface elements include a layer/legend control, properties panel, toolbar, and status bar. Commands can be defined within the Web layout that call JavaScript functions or make requests to server-side script.
PrintLayout	Defines a layout of a page for printing a view of a map, including scale bar, north arrow, legend, title, and user defined text and graphics.
Folder	Defines a folder for the hierarchical storage of resources.

Parcels.LayerDefinition. In MapGuide this is called a resource identifier and it is used throughout the software to refer to resource documents.

Two distinct types of resource storage or repositories exist in the Resource Service. The Library repository is shared by all users and stores resource documents for an indefinite period of time. The Session repository stores transient resources that exist only for the lifetime of a client session. A unique session repository is created for each client session and only that client may access it. When the client session ends, that session's repository is deleted along with its content.

7.4.2 Feature Service – Spatial Data Access

The Feature Service provides a common set of operations for describing, querying, updating, and analyzing spatial data regardless of where or how they are stored. More specifically, the Feature Service provides the ability to interrogate the schema of a data store, query data using spatial and/or attribute filters, perform aggregate queries, update spatial or attribute data, insert new features, and delete features (see Fig. 7.3).

The Feature Service makes heavy use of another OS project called Feature Data Objects (FDO) (http://fdo.osgeo.org). FDO is an API for manipulating, defining and analyzing geospatial information that is completely data store agnostic. FDO uses a provider-based model for supporting a variety of spatial data sources, where each provider typically supports a particular data format or data store. The logical data model exposed by FDO is an extension of the Open Geospatial Consortium's (OGC) Simple Features standard.

The list of available FDO providers is growing on a regular basis. At the time of writing, those included with the MapGuide OS distribution or available for download are shown in Table 7.2. More will likely be added over time.

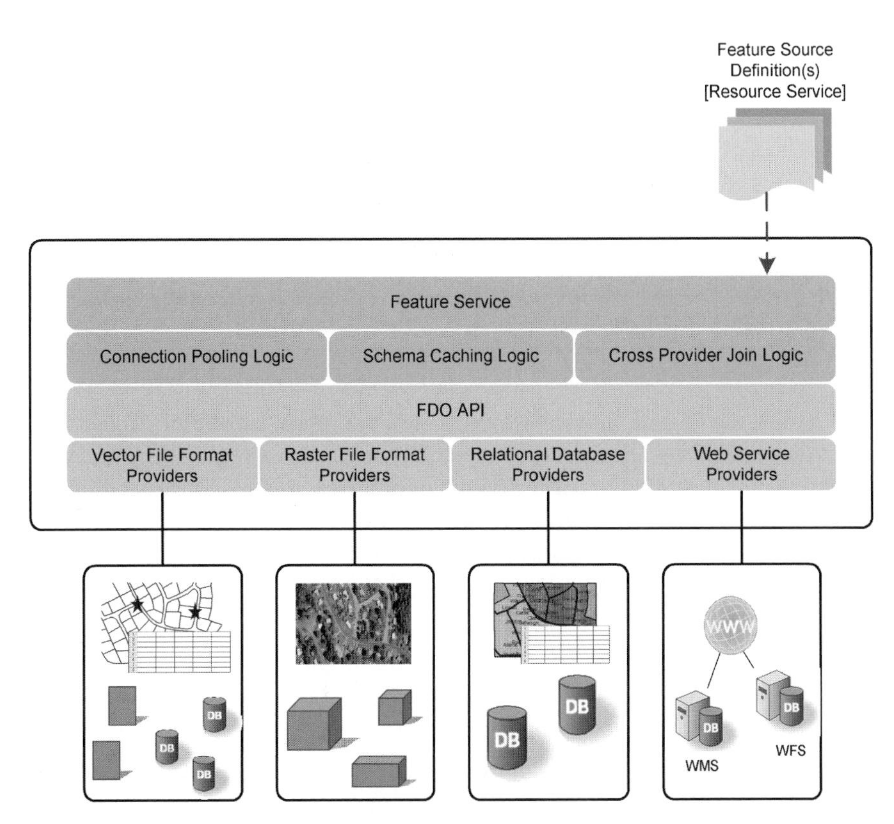

Fig. 7.3 MapGuide feature service

Table 7.2 MapGuide FDO providers

Provider	Description
SDF	Spatial Data Format (SDF) is the native file format of FDO. The FDO provider for SDF supports multiple feature classes per file with a schema defined number of data and geometric properties per feature class. The SDF format is optimized for fast spatial reading of data in single writer multiple reader scenarios.
ESRI SHP	The FDO provider for SDF supports native read/write access to ESRI shape (SHP) files.
OGR	The FDO Provider for OGR provides read-only access to a variety of vector file formats supported by the open source OGR library.
GDAL	The FDO Provider for GDAL provides read-only access to a variety of raster file formats supported by the open source GDAL library. Directories of correlated raster files are automatically treated as a single uniform set of tiled images.
ESRI ArcSDE	The FDO Provider for ArcSDE supports full read/write access to ESRI ArcSDE 9.1 and 9.2. The provider exposes most of the capabilities of ArcSDE including full transactional support and data versioning.
PostGIS	The FDO Provider for PostGIS supports full read/write access to PostGIS database systems.
MySQL	The FDO Provider for MySQL supports full read/write access to MySQL database systems.
Oracle Spatial	The FDO Provider for Oracle supports full read/write access to Oracle Spatial and Locator database systems.
SQL Server	The FDO Provider for SQL Server supports full read/write access to Microsoft SQL Server and adds spatial extensions on top of the database's native capabilities.
ODBC	The FDO Provider for ODBC supports access to simple point feature and attribute data. Point features are exposed by mapping x, y, z columns to geometric properties.
OGC WMS	The FDO Provider for WMS supports access to OGC-compliant Web Map Services.
OGC WFS	The FDO Provider for WFS supports read-only access to OGC-compliant Web Feature Services.

In Fig. 7.3, the Feature Service provides a few additional capabilities above and beyond those of FDO. These include the ability to pool and reuse connections to FDO Providers, the ability to cache schema information so that it only needs to be queried from data stores once, and the ability to establish 1:1 and 1 : many joins across FDO Providers. Defining a join in Feature Service effectively creates an extended feature class that can subsequently be used as a source of data for a map layer.

7.4.3 Drawing Service – Computer Assisted Design Data Support

The Drawing Service provides read-only access to the content of an Autodesk DWF file. The use of DWF allows MapGuide to incorporate computer assisted design

(CAD) content while preserving the visual fidelity of the CAD application. This is possible because DWF files contain pre-rendered graphics and metadata. For example, the geometric elements already have line colour, pattern, fills, and other styles defined. DWF is an open format, and Autodesk offers a publicly available source code library for reading and writing DWF files.

7.4.4 The Mapping, Rendering, and KML Services – Map Output

The Mapping, Rendering, and Keyhole Markup Language (KML) Services provide operations for producing map output, including generation of legend elements. Each of these services is responsible for a different type of map output.

The Mapping Service produces either static or dynamic vector-based DWF output for use with the freely available Autodesk DWF Viewer or Autodesk Design Review software. Static or EPlot DWFs can be produced for use offline or for high fidelity map printing and plotting. Dynamic or EMap DWF retains its connection to the MapGuide Server, and the Autodesk DWF Viewer intelligently uses that connection to refresh content as the user pans, zooms, and interacts with the map.

The Rendering Service produces map output as images in a variety of formats including PNG, JPG, and GIF. Operations are included for producing a map as a single image and for producing a map as a set of tiles with overlay images for dynamic content and selection rendering. To support selection and map interaction in thin clients that are based on displaying map images, the Rendering Service also contains operations to query features on the map that meet a specified spatial criteria, such as all features at a specified point, all features that intersect a specified bounding polygon, etc.

The KML Service produces map output as a stream of KML or zipped KML (KMZ) code. KML is the XML format understood and consumed by Google Earth. Operations are included for generating KML for a single layer or an entire map. Wherever possible the KML Service attempts to adhere to the stylization rules defined in the Layer Definition resources when producing KML, although it is not a perfect match because KML does not support complex symbology such as custom line patterns, or hatch patterns.

All three of these services rely on a common stylization component, as shown in Fig. 7.4. The MapGuide Stylizer takes input from the calling service operation, reads each layer definition using a Resource Service, queries the feature or drawing data, performs coordinate transformation if necessary, and then applies the stylization rules using the appropriate renderer component to produce the desired output format. By having a common MapGuide Stylizer component, new output types and formats can be added to MapGuide without any architectural or structural changes to the MapGuide source code.

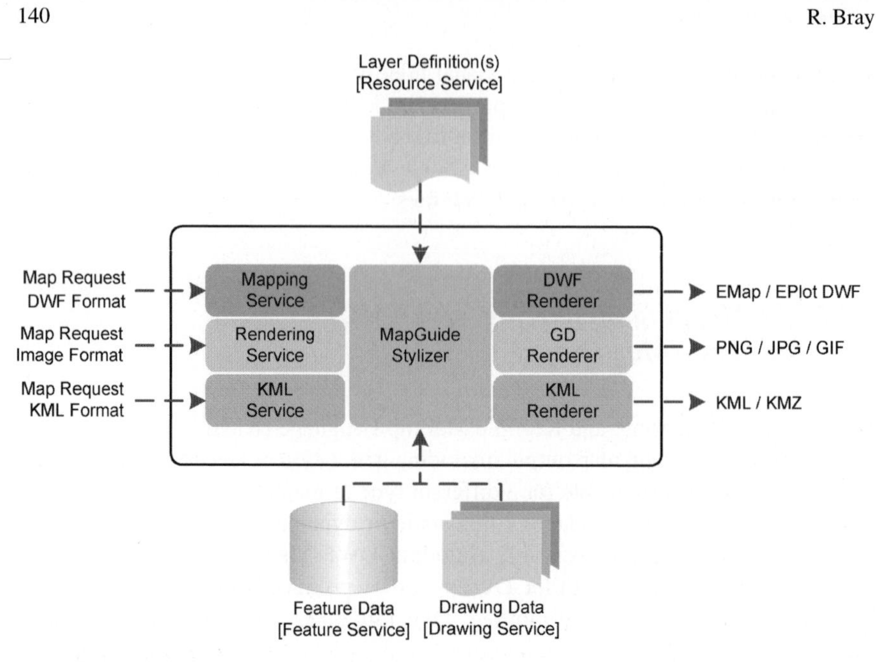

Fig. 7.4 MapGuide stylizer

7.4.5 Tile Service – Caching Map Image Output

The Tile Service supports the concept of "tiled" maps. The use of "tiled" maps is popular in AJAX clients where maps are displayed as a matrix of map images to improve performance and user interaction. The Tile Service provides operations for retrieving cached map image tiles and for managing the tile cache. The Rendering Service first looks in the tile cache when it is asked to render a map image tile. If the tile already exists it is returned by the Rendering Service, otherwise the tile is rendered and stored in the cache first, then returned by the Rendering Service.

7.5 MapGuide Web Tier

The MapGuide Web tier consists of one or more Web servers configured to run the MapGuide Web Server Extensions. The Web Server Extensions play two very important roles. First, they expose a set of operations over HTTP that can be used directly by Web-based clients such as the MapGuide Viewers, MapGuide Web Studio, and Google Earth. Second, they provide a rich API that can be used to build Web-based GIS applications.

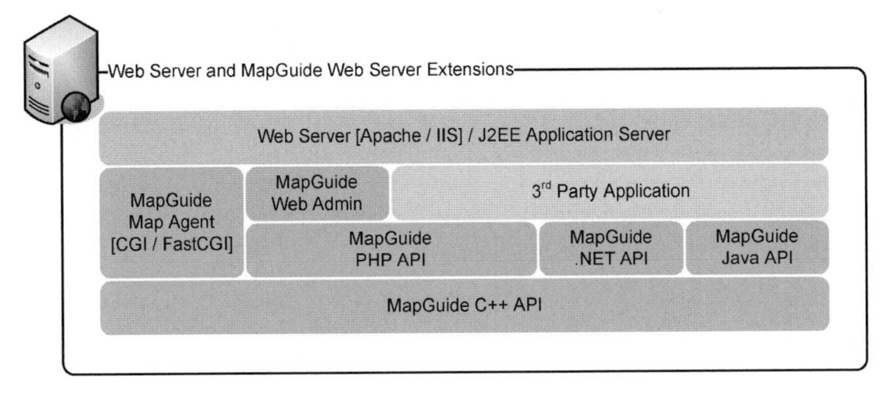

Fig. 7.5 MapGuide web components

The components that make up the MapGuide Web Server Extensions are shown in Fig. 7.5. The MapGuide Map Agent plugs into the Web server as a common gateway interface (CGI) or the OS FastCGI extension module. The Map Agent intercepts HTTP requests from clients and processes them using the MapGuide APIs which may in turn require the assistance of a MapGuide Server.

The CGI operations are used extensively by the MapGuide Viewers and Web Studio to perform actions requested by the user, such as rendering maps and configuring resource documents. In addition, the Open Geospatial Consortium (OGC) Web Map Service (WMS) and Web Feature Service (WFS) request encodings are understood by the Map Agent and implemented using the MapGuide APIs. Many of the CGI operations can be directly called using a simple set of test forms available in MapGuide, which can be accessed via the following URL:

http://<Webservername>:<Webserverport>/mapguide/mapagent/index.html

As noted earlier, the MapGuide APIs are implemented in C++ and exposed to application developers in the popular Web development languages PHP, .NET, and Java. Web application developers can use these APIs to create sophisticated applications that query and update feature data, perform spatial analysis that generates new temporary or permanent layers, and display those layers as part of the users map. Many of the default MapGuide Viewer commands such as Buffer and Select Within are written as applications that make use of the MapGuide API.

One final component of the MapGuide Web Server Extensions is the Web-based Site Administration application. This application, written entirely in PHP, is shown in Fig. 7.6. It enables a system administrator to configure MapGuide Servers remotely, configure and view server logs, manage users and groups, create and load data packages of resource content, configure OGC service properties, and define logical aliases for physical data directories. The MapGuide Site Administrator application can be launched from the following URL:

http://<Webservername>:<Webserverport>/mapguide/mapadmin/login.php

Fig. 7.6 MapGuide site administrator

7.6 MapGuide Client Tier

The MapGuide client tier consists of Web browsers, Google Earth, or any client software that supports the OGC WMS or WFS standards. In the future, MapGuide may play the role of a geospatial middle-tier supporting either thin client or thick client editing applications.

The MapGuide Viewers run inside a Web browser and can make use of either client AJAX technology or the plug-in oriented Autodesk DWF Viewer. The current MapGuide Viewer is based on frames, where the content and visibility of the frames are defined by a resource document known as a Web Layout. The structure of the MapGuide Viewer frameset is shown in Fig. 7.7. The highlighted frames contain visible user interface elements. All other frames are structural and invisible to the user.

The content of the mapFrame depends upon whether the AJAX or DWF form of the viewer is in use. For the AJAX Viewer the mapFrame contains HTML elements representing the map, legend, and property pane. For the DWF Viewer the mapFrame contains an embedded ActiveX Control. Because the DWF Viewer is ActiveX based, it only works in Microsoft's Internet Explorer browser running on Microsoft Windows. The AJAX Viewer on the other hand runs in all popular Web

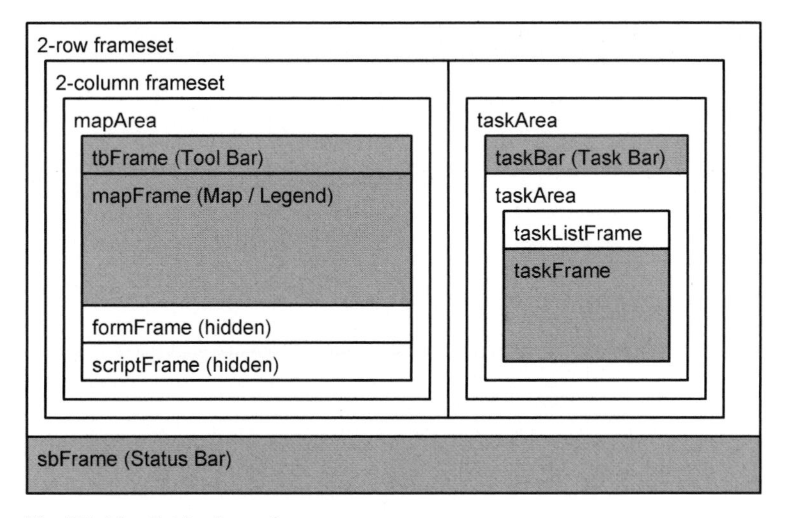

Fig. 7.7 MapGuide viewer frame structure

browsers including Internet Explorer, FireFox, and Safari. Generally speaking, the two different forms of the MapGuide Viewer support the same set of capabilities.

Both Viewers support all of the common map navigation commands, single and multi-feature selection, display of selected feature properties, map tips that display feature information when the user hovers the mouse, printing, and more advanced commands that support buffering, selection of features that intersect the bounds of another, and some rudimentary search capabilities. Generally speaking, however, the AJAX Viewer is the preferred choice for most applications. Besides supporting multiple browsers and platforms, it also supports tile-based maps with smooth pan and zoom capabilities often referred to as "slippy maps". The DWF Viewer is sometimes a compelling choice because selection feedback and property display is more responsive than the AJAX Viewer which has to send requests to the server to perform those operations. Figure 7.8 shows an example of the MapGuide AJAX Viewer with a default tool bar, status bar, and legend/property pane. The map view shown in this figure performs basically the same generic functions as other viewers from both the OS and proprietary alternatives that are available.

The frames-based nature of the MapGuide Viewer is often seen as outdated and inflexible. However a more modern and flexible viewer technology will be used in new releases. Once this is fully implemented, new applications will be migrated to it. However, the number of legacy frames-based applications will see this form of presentation also remaining in use for some time. This is discussed in Sect. 7.8 where future directions for the MapGuide project are summarized.

Google Earth can also be used as a MapGuide client, either to view two dimensional (2D) data in a broader context, or to display 2D data extruded into three dimensions (3D). An example of this can be seen in Fig. 7.9, which shows the building footprints for downtown Chicago from an ESRI SHP file extruded and displayed in 3D. Any MapGuide Map Definition resource can be displayed in Google Earth via

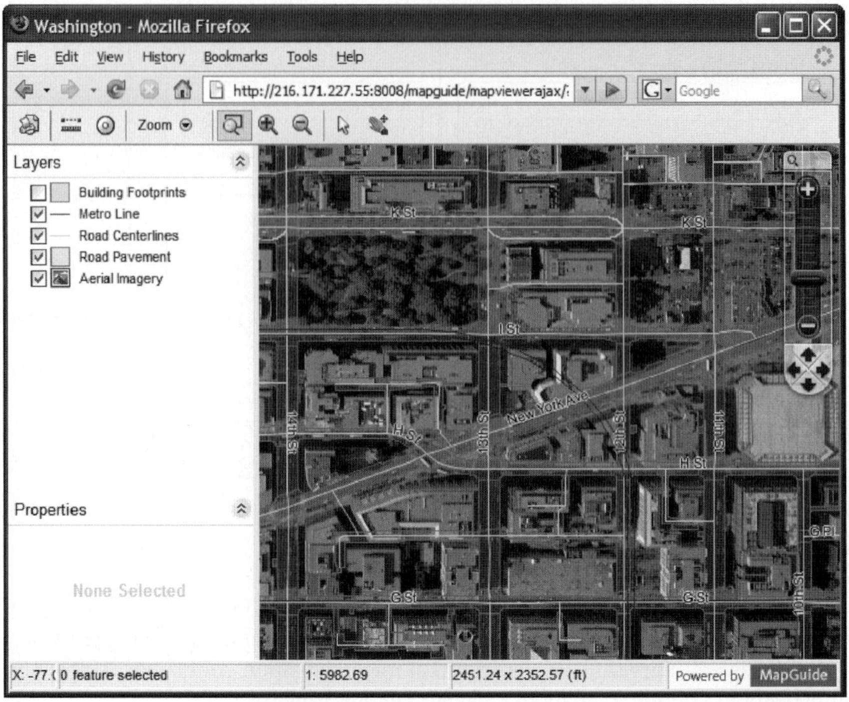

Fig. 7.8 An example of the MapGuide AJAX viewer

the following URL, which may be added as a Network Link directly in Google Earth or from a Web Browser:

http://<*Webservername*>:<*Webserverport*>/mapguide/mapagent/mapagent.exe?
OPERATION = GetMapKml&
VERSION=1.0.0&
USERNAME=Anonymous&
PASSWORD=&
MAPDEFINITION=
Library://<*resourcepath*>/<*mapname*>.MapDefinition&
FORMAT=KMZ

By using the MapGuide APIs, it is possible to extend the Google Earth support by adding polygon centroid features, links to informational reports, photos, and many more functions. Because of the way Google Earth supports hyperlinks and Web pages there are very few limits to what can be done.

Another bundled client is MapGuide Web Studio, a Web-based application for defining resource content such as Feature Sources, Layer Definitions, and Map Definitions. When a user connects to a MapGuide Server with Web Studio, a tree-like structure that shows all of the resources managed by that server is presented on the upper left side of the window. As shown in Fig. 7.10, a list of resource documents

Fig. 7.9 Using Google Earth as a MapGuide client

that are open for editing, one of which will be highlighted, are presented on the lower left. The open resource document appears on the right hand side of the window as a number of panels. These panels allow the user to edit easily various aspects of the resource document through a forms-based interface. The bottom-most panel on the right hand side of the window allows the user to preview the changes that are being made to the resource.

Use of this approach to Web map document presentation provides a simple and highly efficient means of rendering map presentations within the viewer using familiar Windows-based interface mechanisms such as a Windows Explorer-like navigation pane and form fields. This removes the need to edit text in a script-based file in order to render a map image for final appearance in a Web-based browser.

7.7 Developing Web Applications with MapGuide

Developing MapGuide Web applications is typically a two-step process. The first step involves defining all of the spatial content including data sources, map layers, one or more maps, and a Web layout. The second step involves customizing and

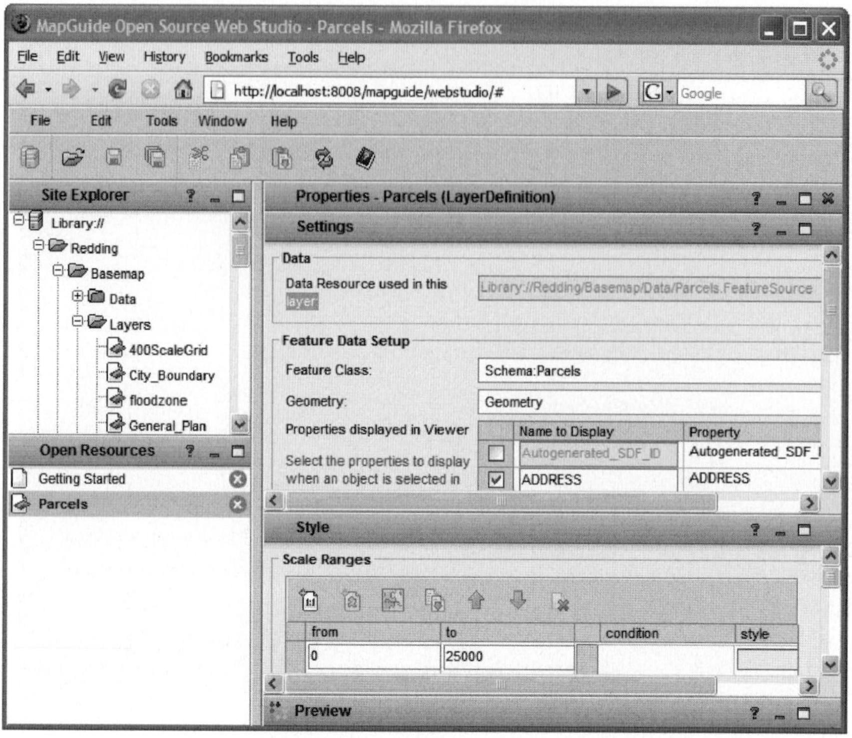

Fig. 7.10 Editing a layer in MapGuide web studio

extending the default viewer with the features and functionality required by the end users of the application.

7.7.1 Defining an Application's Resources

The first step toward creating a MapGuide application is defining all of the geospatial content, as shown in Fig. 7.11, and storing that content using the Resource Service. These steps are usually done with an authoring application such as MapGuide Web Studio noted above, but can also be done manually without the use of a graphical user interface.

The first step in the workflow involves defining sources of feature and drawing data. Feature Sources define the locations of data files or connections to databases, while Drawing Sources define the locations of CAD data in DWF format. Once the sources of data are defined, Layer Definitions can be created that specify the source of the data for the layer via a reference to a feature or drawing source, what scale ranges the data should be displayed at, and how to stylize the data at each scale range. Stylization can either be static or rule driven to produce thematic maps. In

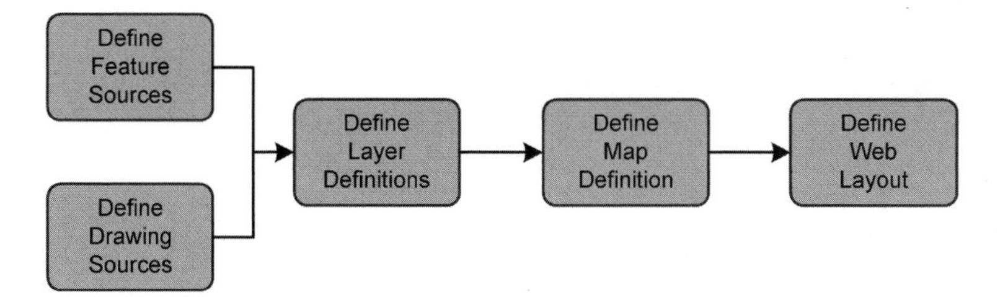

Fig. 7.11 Defining an application's resources

both cases automatic labelling rules can also be set up on a per scale range and per thematic rule basis.

Once the Layer Definition resources are defined, they can be combined into a map. The Map Definition resource specifies the extent of the map, the coordinate system to be used for rendering the map, a background color if desired, and an ordered list of Layer Definition references. The order of the layers determines the order in which they are drawn. For each layer, the legend label, default visibility, and selection behaviour can be specified. Layers may also be logically grouped into a hierarchy if desired.

The final resource type in the workflow is the Web Layout. Web Layout resources contain a reference to a Map Definition, some basic visibility controls that allow various user interface elements to be turned on or off, tool bar and context menu definitions, and command specifications. Commands in a Web Layout can be based on well known tools, such as Pan, Zoom In, and Select, or commands can invoke client- or server-side code written by an application developer. Once the Web Layout resource is defined it can be viewed in a browser via the following URL:

http://<*Webservername*>:<*Webserverport*>/mapguide/mapviewerajax/?
WEBLAYOUT= Library://<*resourcepath*>/<*layoutname*>.WebLayout&
USERNAME=Anonymous&
PASSWORD=&
LOCALE=en

7.7.2 Programming MapGuide Applications

MapGuide's API offers extensive opportunities to customize and extend the default viewer with the features and functionality required by the end users of the application. The APIs are available in PHP, Microsoft .NET, and Java, and are identical in all three environments.

The basic structure of a MapGuide application is shown in Fig. 7.12. In the Web Layout the application developer can create custom commands that invoke server-side scripts written as PHP, ASP.NET, or JSP pages. With the invocation of these commands MapGuide can pass along information about a map's state, such as the

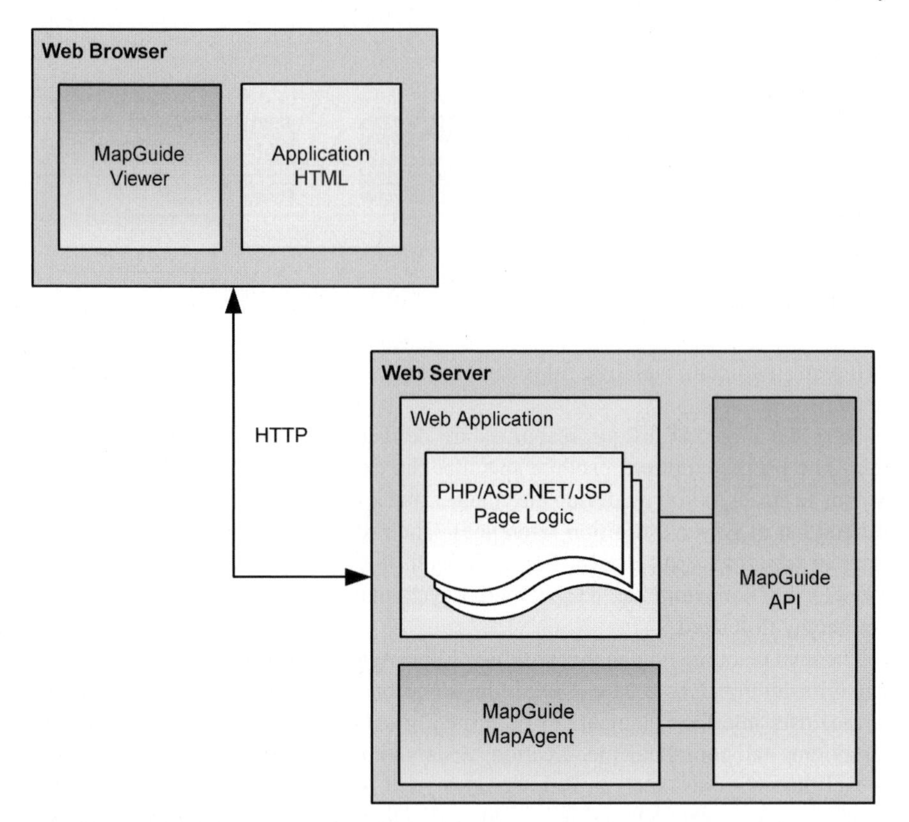

Fig. 7.12 MapGuide application structure

current list of selected features. Using the MapGuide APIs, the server-side script can query and update feature data, perform spatial analysis, create temporary feature data and layers, and otherwise manipulate the state of the map.

The MapGuide APIs are packaged into three groups as shown in Fig. 7.13. The System APIs define basic data types, collections, and exceptions that enable MapGuide to support multiple programming languages in a uniform and consistent manner.

The Runtime APIs execute locally in the Web tier. The Geometry APIs are based on the GEOS library (http://geos.refractions.net) which provides a very good implementation of the OGC specification for simple features, including spatial relationship tests, overlay functions, and a buffer operation. The Coordinate System libraries support coordinate re-projection, linear and great circle distance measurement, and other related operations. The Feature Schema APIs provide the ability to interrogate the schema of a feature source, including the feature classes, properties, data types, and relationships. The final set of Runtime Map and Layer Manipulation APIs enable the application developer to interrogate and modify the state of the

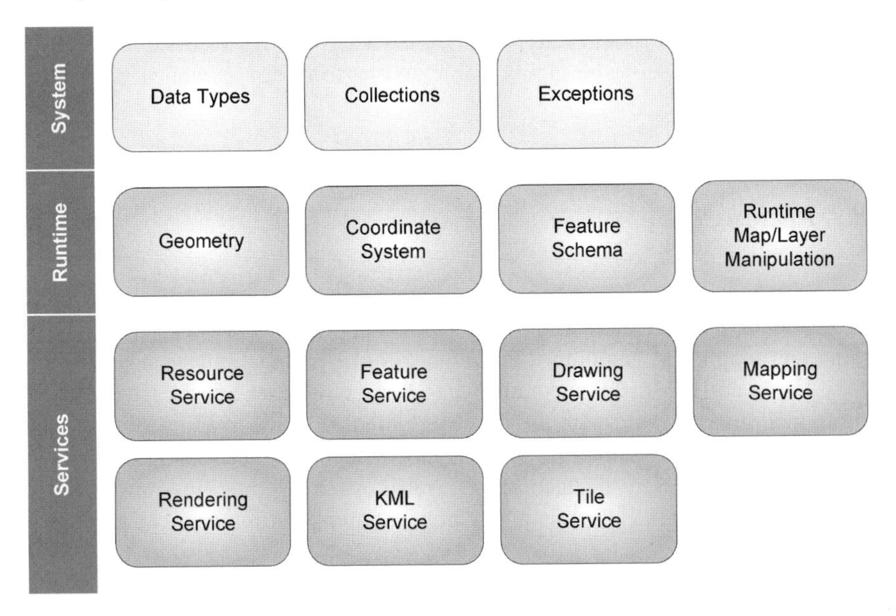

Fig. 7.13 API packages

map session. This includes the ability to show or hide layers, add or remove layers, modify the definition of a layer, control the current map extent, map centre, and map scale.

The Service APIs execute in the server tier and expose all of the functionality of the MapGuide Services, including resource content management, data query and update, and map rendering operations.

7.8 Future Directions

Development on the MapGuide Open Source project continues, and new features are introduced on a regular basis. Constant evolution is one of the reasons users are drawn to open source projects. This section highlights some upcoming features as well as some longer term initiatives.

7.8.1 Replacing FastCGI with an Apache Module and ISAPI Extension

The MapGuide project has had good success with FastCGI technology as a means to improve upon the performance of CGI. However, the overall stability of FastCGI is very problematic. In the near future the FastCGI MapAgent will be replaced with an

Apache Module and an ISAPI Extension (for the Microsoft IIS Web Server). These technologies are proven to be highly stable and perform just as well if not better than FastCGI.

7.8.2 Fusion

A more modern and flexible AJAX viewer framework code named Fusion is currently under active development. Fusion will use the Open Layers project for map display and map event handling (http://www.openlayers.org). Fusion page layout will be based on HTML div tags rather than frames and user interface widgets or controls will be associated with div tags by reference, with cascading style sheets defining the visual appearance of the widgets. A new Application Definition resource type will be used in place of the Web Layout resource type to associate widgets with div tags in HTML template files. A sample Fusion application can be seen in Fig. 7.14.

Although this differs only slightly in appearance from Fig. 7.8, the architecture is substantially different.

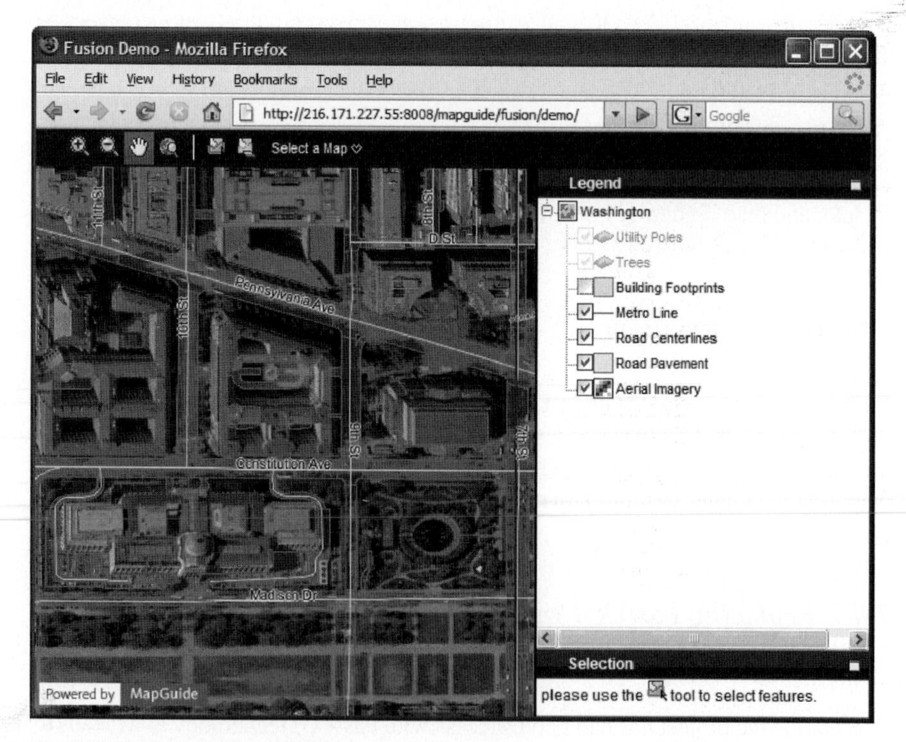

Fig. 7.14 Fusion sample application

7.8.3 Additional Standard Support

OGC standards are becoming increasingly more powerful and important in the spatial information technology industry, particularly in federal and local government. This is primarily because open standards drive interoperability across organizations, products, and vendors. As noted earlier, MapGuide supports the Web Map Service (WMS) and Web Feature Service (WFS) standards, but needs to stay current with the OGC standards and implement some additional ones like WFS Transactional (WFS/T), Web Coverage Service (WCS), and Web Map Context (WMC) in particular.

7.8.4 Web Services

Web Services are becoming an increasingly pervasive and compelling technology for constructing enterprise applications. The MapGuide architecture is an ideal platform for exposing a set of spatial Web services based on either the Service Oriented Architecture Protocol (SOAP) or the Representational State Transfer (REST) architectural style. Initially, these interfaces would logically expose the functionality of the existing MapGuide Services. However, it is also possible to imagine web service interfaces for performing coordinate conversion, spatial analysis, and more.

7.8.5 Metadata Catalog

The Resource Service currently manages a large database of geospatial content. Extending the Resource Service to support associating International Organization for Standardization (ISO), Federal Geographic Data Committee (FGDC), or OGC metadata with each resource is a logical step. Add to that a Catalog Service, and MapGuide would be able to advertize its spatial content and allow rich metadata search capabilities. In many ways this is an extension of standards support and dovetails nicely with the recommendations in Sect. 7.3.

7.8.6 3D Rendering/Viewer

With the availability of Google Earth, Microsoft Virtual Earth, and new data capture techniques such as Light Detection and Ranging (LIDAR), three dimensional (3D) mapping is becoming much more commonplace. As this trend continues, MapGuide will need to make the move to embrace 3D completely, both in terms of visualization and analysis. On the analysis side there are a number of interesting algorithms that are useful to geospatial data, such as line of sight, that can be visualized two dimensionally but have a much greater visual impact in 3D.

7.9 Conclusion

MapGuide Open Source is a young but healthy and growing free and open source for geospatial (FOSS4G) project. It provides powerful Web-based spatial processing and rendering capabilities that typically are only found in desktop GIS software or commercial server software with hefty license fees. Because of this fact, its user community is expanding rapidly. In addition, the MapGuide Open Source project is backed by a commercial software vendor, Autodesk, Inc, which offers a low cost commercial version of the software in parallel. Autodesk made the choice to release MapGuide as an open source project because they felt that web mapping technology was becoming a commodity, and their business opportunities were better served by focusing on solutions built on its core technology. Because Autodesk depends on the core Web mapping technology for their solutions, they remain committed to ensuring the ongoing success of the MapGuide Open Source project. To learn more about the MapGuide Open Source project, try out the software, and/or join the community visit http://mapguide.osgeo.org.

Chapter 8
GeoTools

Ian Turton

Abstract This chapter introduces the GeoTools library, which is a Java GIS toolkit. The aim of the library is to allow programmers who are developing geospatial applications to concentrate on building the interesting parts of an application, while reusing generic tools for basic functions, such as reading an ESRI shapefile, or styling a set of features using a styled layer descriptor file. GeoTools provides these basic underlying components that virtually all spatial data processing applications need, but that would be prohibitively expensive in terms of time and effort for each project to develop independently. The history of the GeoTools library and a brief overview of its main components are given, followed by an outline of the future development paths of the library.

8.1 Introduction

GeoTools is a Java library for geographic information system (GIS) applications. It is now in its second version and is used as the base of several well known free and open source for geospatial (FOSS4G) products including GeoServer, uDig, and GeoVISTA *studio*. By building on top of other open source (OS) libraries, such as the Java Topology Suite (Vivid Solutions) and the Open Geospatial Consortium's (OGC) GeoAPI, GeoTools leverages the best of these lower level libraries to provide a mid-level library of functions that simplifies the construction of complex spatial data processing applications, while hiding the complexities of data sources, feature models, and projections from the end user.

Ian Turton
GeoVISTA Center, Pennsylvania State University, University Park, PA 16802, USA,
e-mail: ijt1@psu.edu

G.B. Hall, M.G. Leahy (eds.), *Open Source Approaches in Spatial Data Handling*.
Advances in Geographic Information Science 2,

8.1.1 Users

GeoTools is designed as a library to be used by programmers building spatial data applications, rather than by end users looking to make maps. Of course, in many cases developers may also have the ambition to draw maps, but they are often prepared to go to great lengths to avoid doing this when a more general way can be found.

The first version of the GeoTools library had as a primary goal to allow users to draw maps in a simple and interactive manner. However, this led to the project acquiring more users than developers as people started to use the code and expected it to work and be improved without any clear knowledge of the amount of work involved in supporting enhancements. This situation was solved by starting a second version of the library which was much harder to use and had far fewer user interface components. This change in approach meant that the project tended to focus on developers, who were happier with the concept of having to write more of their own code to produce a useful program, rather than general users, although end use clearly remained a basic goal of the effort. It is likely that this change in focus by the early developers is what saved GeoTools from collapsing under the weight of user demands, which often ends new and promising OS projects before they reach their full potential.

8.2 History

Sun released the first public version of Java in 1995, the same year that the initial developer of GeoTools, James Macgill, started a master's degree in GIS at the University of Leeds in the United Kingdom. By 1996, it was apparent that the release of Java was going to change the face of interactive Web browsing. By this time, the platform was also being used by Macgill to visualize results in spatial cluster detection (Macgill 2001). As an increasing number of projects in the newly formed Centre for Computational Geography (CCG) at the University of Leeds needed custom mapping, the simple viewer that Macgill had built for the purposes of his doctoral research was expanded and the GeoTools project was born.

8.2.1 Visualizing Spatial Cluster Detection

One of the initial areas of interest in the CCG's research was spatial cluster detection. This is the process where a program is run to determine whether the distribution of a specific form of cancer, for example, is clustered in space or not. While early methods like the geographic analysis machine (GAM) (Openshaw et al. 1999) were relatively easy to use and debug, more complex methods based on flocking boids (Macgill and Openshaw 1998) were being investigated (Fig. 8.1). This led to Macgill developing a basic mapping system in Java that could read and display

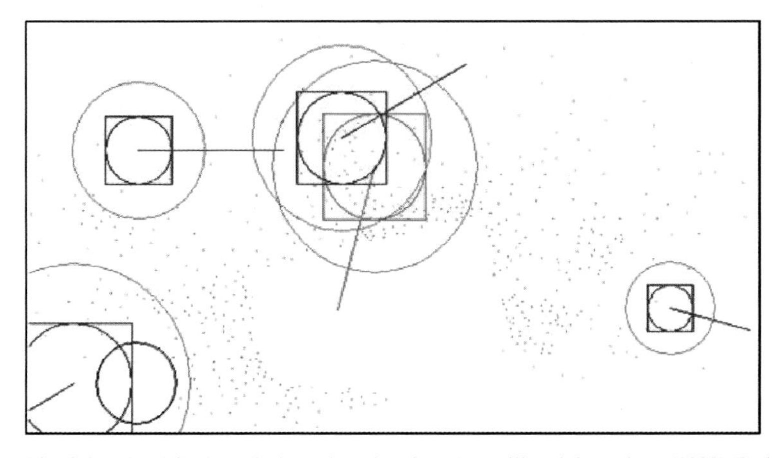

Fig. 8.1 A boid flock exploring crime data from Macgill and Openshaw (1998) Each point represents the population and crime count for a city block. If the crime count is greater than one it is coloured differently

Environmental Systems Research Institute (ESRI) shapefiles in a geographically correct manner, while also being able to move animated "sprites" over the map surface quickly and easily. In this case, a sprite refers to a two dimensional image or animation, which is integrated into a larger computer-generated scene.

8.2.2 Virtual Slaithwaite

At the same time that Macgill and Oppenshaw were working on spatial cluster visualization, other researchers in the CCG were working on a public participation GIS (PPGIS) project that required the ability to publish interactive maps on the Web. In June 1998 the West Yorkshire Village of Slaithwaite carried out a "Planning For Real"® (PFR) exercise to identify the views and opinions of local residents regarding the local environment and how they would like to see their village develop in the future. It was clear that to implement this system it would be much more expeditious to expand Macgill's existing mapping code than to develop an entirely new mapping application. Hence, the extant code contained in GeoTools became a library for a group of users to build upon, rather than just a research tool that was focused on a relatively limited problem domain (i.e., spatial visualization on the Web).

The Virtual Slaithwaite system (Kingston et al. 2000) was an on-line GIS tool and was arguably among the first such systems available to the public to facilitate a two-way flow of information. The Web browser window consisted of four frames each containing particular pieces of information (see Fig. 8.2) related to the objective of the application (gathering and displaying participatory input). Members of the public could view a map of the town Slaithwaite, perform zoom and pan operations to assist in visualization and navigation, select features and ask questions such as "what is this building?" and "what is this road?", and then make suggestions

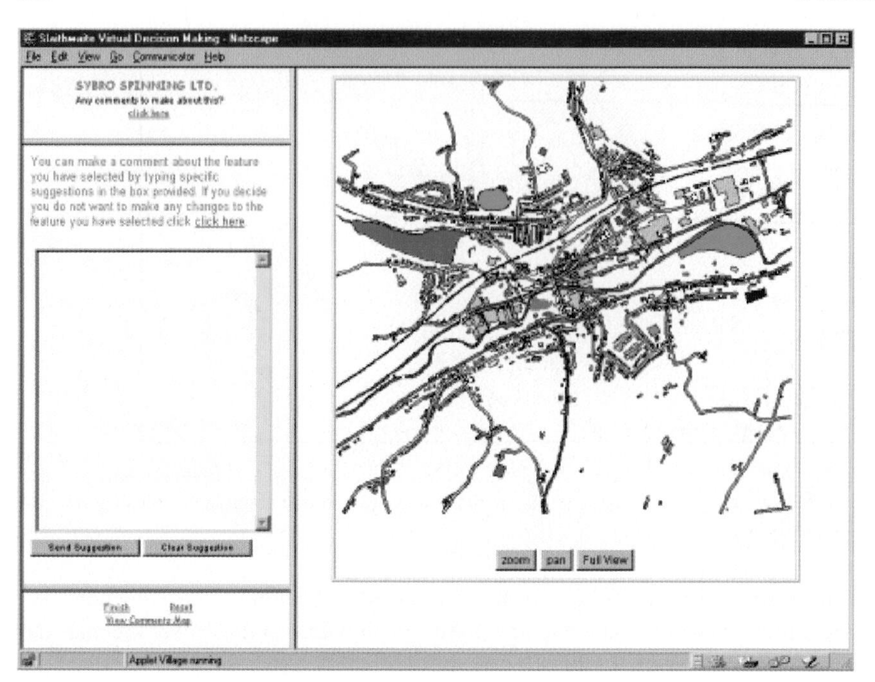

Fig. 8.2 The Virtual Slaithwaite Web-based system

about specific features identified from the map. All user input was stored in the Web access logs, which were then used for future analysis and feedback into the planning process. In this manner a community database was created, representing the range of views and feeling about planning issues in the village.

The left hand side of the screen in Fig. 8.2 displays the current selection in the top frame. When a building is selected the middle frame is populated with a form allowing the input of text relating to that building. For other features, such as open spaces, the river, or the canal, a free form text box is provided allowing the user to type whatever they wish about the selected feature. This information is then committed to a database for future analysis, not by the public but by researchers working in tandem with village planners. When the user finishes their selections and comments, they exit the system and are provided with a series of questions seeking feedback on their use of the tool for the developers to make refinements to the system. This same basic concept underlies many other participatory Web mapping tools (see Chap. 11), which have grown in their sophistication since the Slaithwaite project.

8.2.3 GeoServer

One of the first external projects to make use of the GeoTools library was GeoServer. GeoServer was started by The Open Planning Project (TOPP), a small non-profit organization in New York City looking to make urban planning decisions more open to citizens. TOPP's objective was to create a suite of tools to enable open democracy in planning and to help make government more transparent. Hence, the use of mapping tools must be transparent and modifiable so that they do not surreptitiously put forth an agenda that supports the views of any one group over another. The tools must also be freely available so that the public at large can participate, rather than only those who can afford to spend money on purchasing software. For this reason all tools in the GeoServer suite must be OS and freely available.

After wrestling with the fact that an internal model of geographic features was needed, and not wanting to replicate a great deal of work, GeoServer discovered the GeoTools project at the time it was embarking on the Virtual Slaithwaite Project and some early experiments were undertaken with the emergent OGC WMS standard (then known as the Web Mapping Testbed). This gave GeoTools an exciting opportunity to support real users, who were not the developers, in use of the software. This led to an explosion of new code, and soon led to the discovery of some fundamental problems with the way features were handled in the existing GeoTools environment.

8.2.4 GeoTools2

When the GeoTools project met the GeoServer project, it was apparent that there was a need for a better feature model. In fact, as programmers outside of the core developer group joined the unified project, it became increasingly clear that much of the GeoTools library could be improved by using the lessons of the previous four years. It was decided that work should be halted on the existing library and a new version should be built from the ground up to be known as GeoTools2. In 2002, version 2.0 of GeoTools was formally released on the geotools.org website (http://www.geotools.org). The development team had expanded from the original developers to include an international group which also formed a project management committee responsible for overall control of the project. Interestingly, the first version of GeoTools never actually made it to a formal 1.0 release. It was finally abandoned at release 0.8, in part due to the fact that the developers felt people might expect it to be robust if it left beta.

The aim of the rebuild was to make GeoTools a more modular and robust library which other projects and developers could make use of. The addition of the SEAGIS project (http://seagis.sourceforge.net/), developed by Martin Desruisseaux, to the GeoTools project's code base in 2002 provided coordinate transformation services, grid coverage, and rendering implementations, and gave the fledgling project a huge boost.

8.2.5 Organization

As GeoTools grew from a project maintained by one developer to one with ten developers, it became apparent that an organizational structure was required. In common with many open source projects, GeoTools developed from the well known "benevolent dictator" model, where one person has absolute control, to a more open management committee drawn from the active developer base. To be considered a developer, a programmer has to have contributed useful and working code to the project on several occasions (either as patches to existing parts of the project or as new functionality). When new code is positively reviewed by existing developers, a proposal is made to give the new developer write access to the GeoTools code repository so that he/she can add code directly to the project's code base. This proposal is voted on by the existing members and requires at least three positive votes and no negative votes (votes can be $+1$, 0 or -1) to proceed. A step up from developer within the GeoTools organization is a module maintainer, which is a developer who has overall responsibility for a whole module within the GeoTools environment.

While much of the day-to-day development of GeoTools can be carried out in a decentralized manner with one module unaffected by changes in another, some changes need to be coordinated across the library as a whole. This is the task of the project management committee (PMC), which meets once a week using Internet Relay Chat (IRC) to discuss any issues of interest. All developers and users are welcome to the meeting, but PMC members are expected to participate. The time of the meeting changes occasionally to accommodate changes in PMC membership, but details can always be found at http://docs.codehaus.org/display/GEOT/3.1+Internet+ Relay+Chat. Other day-to-day issues of GeoTools development are discussed on the project mailing lists (see http://docs.codehaus.org/display/GEOT/3.2+Email for details).

8.3 Openness

From the beginning of the GeoTools project, the developers considered the source code to be something that should be freely shared and developed. GeoTools was born in a university environment and, as universities in Britain in the 1990s were quite closely focused on commercialization of research, the actual licensing of software had numerous legal and administrative implications. Hence, it was decided to give GeoTools to the open development community in order to allow it to reach its potential without becoming entrapped in red tape.

As scientists developing software to support science, the primary interest was, and continues to be, in producing repeatable outcomes. This was something that was especially difficult to achieve in the emerging field of computational science in general and computational geography in particular during the 1990s. GeoTools was written in part to avoid this problem by allowing everyone to base their experiments on the same code.

The following sections outline three specific types of openness that the developers sought to support in the development of GeoTools, namely open standards, open source, and open science.

8.3.1 Open Standards

An open standard is one that is free for all developers to read and implement. In the geographic world, the most commonly implemented standards are those developed by the Open Geospatial Consortium (OGC), which is a confederation of industry, academic, and government participants who set voluntary internally recognized standards for the geographic Web and other aspects of spatial data.

The GeoTools developers made a decision very early in the development of the library that it would support as many OGC standards as possible. This allows programs written using GeoTools to interoperate with other programs that also implement these standards. The first standards developed by the OGC that GeoTools supported were those related to Web mapping, with the Web Map Service (WMS) specification was initially implemented as a GeoTools project before the code was passed to the GeoServer community.

In more recent years, the GeoTools project has supported the OGC's GeoAPI project, which is attempting to develop a set of Java interfaces to implement application programming interfaces (API) derived from OGC/ISO standards to allow better interoperability between different Java-based geographic projects. The supported standards include the OGC's Topic 1 (ISO 19107), Topic 2 (ISO 19111), Topic 11 (ISO 19115), Grid Coverage Services, OGC GO-1 Application Objects, and Topic 5: OGC Feature Model. This is a more ambitious project, but when completed it should provide great benefits to the whole of the Java-based geographic software and user communities.

8.3.2 Open Source (OS)

OS refers to programming code that is freely available for other programmers to study and improve. This allows developers to identify and correct bugs in code, which can be highly beneficial to the entire community of developers and end users. It also allows others to see how a programmer did something and, if they are so inclined, to make improvements to the code. Most OS licenses ask that developers contribute their changes back to the project so that others can benefit from the fixes or improvements. In the case of GeoTools, this often leads to contributors becoming responsible for the section(s) of code they have developed and being invited to comment on future changes that might affect their contribution(s).

8.3.3 Open Science

As noted earlier, good science has always relied on the ability of others to replicate experiments in order either to prove or disprove reported results. As more science becomes computational in nature, it becomes more difficult to reproduce experiments that another group has carried out. If differences are found between two computational experiments, it is difficult to determine who has the correct result if both groups are working with closed source programs. Differences in results may be introduced by something as trivial as a bug in an algorithm, or converting input data from one format to another so that a different program can be used. However, without consistent and open code, it is difficult to determine the cause of the problem.

By basing experiments on OS libraries it is easier for other groups to use the same code base and to also inspect the code to check that it implements specific algorithms correctly. It was also the hope of the GeoTools developers that, by building a comprehensive library of geospatial functions, it would be possible for computational geography to move beyond the equivalent of a chemist having to blow all their own glass test tubes before every experiment. The researcher's focus may instead be directed towards the few very specialized tools needed for an experiment.

8.4 Design Principles

The first GeoTools library had a core set of design principles beyond the goal of simply solving the problem at hand as quickly and as easily as possible. Over time, these principles have become codified and were extended to set some guidelines for the project, as outlined in the following sections.

8.4.1 Keep It Simple

The basic design principal of GeoTools is, wherever possible, to use the simplest solution for a problem. This does not mean that things are never complex inside the code. In fact, it is often quite the opposite. However, when given a choice, a developer should generally choose the simple option first. Then, if the situation warrants, either because there is a requirement for optimization or because the simpler solution just doesn't work, a more complex solution can be considered and implemented. For example, when the first GeoTools renderer was written to draw maps on the screen, it simply looped through all the features and drew each one sequentially. While this worked, it quickly became clear that this was too slow for interactive rendering of large numbers of features. To overcome this problem, the GeoTools renderer now caches the styled glyph of each feature after the first time it is drawn, thus avoiding the need to restyle the feature the next time it is rendered. This adds a degree of complexity to the renderer pipeline that would have been hard to deal with when it was first developed, but this approach is now needed to produce the rendering speed required for most applications.

8.4.2 Be a Library

To meet the developers' initial goals of aiding the spread of open computational geography, it was necessary that GeoTools could be used easily by developers in their own programs. Thus GeoTools was designed to be a library with a clear API that would hide the underlying implementations. This allowed end users of the library to program against the stable core, while improvements to underlying methods could be made by the library developers as required.

With the start of the GeoAPI project by the OGC, GeoTools was able to pass much of the API definition work to this new sister project and concentrate more on the design of the underlying implementations. However, the two projects share a number of developers, which helps to sort out potential differences between the groups. As noted, GeoTools has chosen to implement standards wherever possible, with particular emphasis on OGC standards. This makes it easier for other developers to integrate their programs that use the same standards.

8.5 Outline of Functionality

A key requirement of being a library is the ability to select only some of the code as it is needed. This has become especially important as the size of the GeoTools code base has grown over time. Currently GeoTools consists of four major modules:

- Main – provides the key functionality of the library, including the feature model, spatial referencing, data handling, and rendering packages.
- Plugins – provides interchangeable parts of the library, which can be added or subtracted as needed by end user programs. This includes all of the actual data source packages and implementations of spatial referencing databases.
- Extensions – provides add-ons to the library that give additional functionality not needed for simple programs, but which might be useful for some users. For example, the Colorbrewer tool (see Chap. 10 and Sect. 8.5.2.2), graph functions, and OpenOffice add-on packages are contained in extensions.
- Demos – provides example and tutorial code to demonstrate the use of the various parts of the GeoTools library.

8.5.1 Core Functionality

The core functionality of the GeoTools library is defined in the main module. This is where the internal model of GeoTools is set up. It is here that the feature model is defined in a way that allows all parts of the program to agree about what a feature is and how it should be represented. This section of the library also defines how to read and write data, how to create filters that can be applied to data streams to select specific features, and how to style a feature in order to render it.

8.5.1.1 Feature Model

The feature model is the heart of the GeoTools library. It was one of the areas of greatest change from GeoTools Version 1 to Version 2. In Version 1 GeoTools had a very simple data model that stored attributes of features as elements in an array to reflect the simple nature of the programs that were being supported. When GeoTools 2 was developed, more complex feature types were envisaged and a more complex feature model was required. A feature is a representation of a geographic object (e.g., a road, river, or house). In GeoTools a feature is stored as an array of Java objects that represent the feature attributes. One attribute is always the geometry (i.e., its representation in space) of the feature. The other attributes can be any type that is required by the feature being modeled. A feature is defined by a Feature-Type object, which represents the schema of the feature. This allows a program to determine what type of object an attribute should be decoded or encoded as when loading or unloading data from features. The feature class provides helper methods that provide the user with a convenient way to work with the attributes. Since an attribute can be any geographic object, features can even contain information about other features, allowing containment to be modeled by the system.

8.5.1.2 Data Stores

GeoTools provides an abstract datastore concept that allows an application developer to access data from a variety of data sources without having to worry about the actual implementation details. For example, data stored in an ESRI shapefile or a PostGIS database is accessed in exactly the same way once the datastore has been opened. The use of a factory design pattern allows developers to pass a request for a new object to the factory class, rather than needing to hardcode the reference to the class. The factory can then look up a specific implementation and return a concrete object that has been determined at runtime to the program. By providing a factory system to access plug-in datastores, GeoTools allows programs to be completely agnostic to data sources, provided that a specific data store has been implemented for the data source required.

The data store interface provides a standardized set of methods to access data. The most commonly used method is getFeatureSource(), which returns a representation of the features in the store. The feature store then allows the user to query the data to determine how many features exist, what the bounding box of the collection is, and to request a collection of features based on a filter. In most cases the underlying implementations provide an optimized version of the operation so, for example, a database will convert the filter into an SQL query and pass that to the database. However, simpler data stores like shapefiles read all the features into memory, and then apply the filter in Java to each feature in turn.

8.5.1.3 Filters

GeoTools implements the OGC filter specification (OGC 2004), which defines three groups of filters:

- Spatial filters, which involve a geometry element and a spatial relationship.
- Comparison filters, which compare attributes of features with other expressions or attributes.
- Logical operators, which allow the joining of other types of filter.

There are 11 spatial filter types defined by the specification:

- Disjoint – True if the two geometries do not touch or intersect.
- Equals – True if the two geometries are the same.
- DWithin – True if the geometry being tested is within the stated distance of the geometry provided.
- Beyond – True if the geometry being tested is beyond the stated distance away from the geometry provided. Functionally equivalent to Not DWithin(...).
- Intersect – True if the two geometries intersect. This is a convenient method, as it is also possible to use Not Disjoint(A,B) to get the same result.
- Touches – True if and only if the only common points of the two geometries are in the union of the boundaries of the geometries.
- Crosses – True if the intersection of the two geometries results in a value whose dimension is less than the geometries, the maximum dimension of the intersection value includes points interior to both the geometries, and the intersection value is not equal to either of the geometries.
- Within – True if the first geometry is wholly inside the second geometry.
- Contains – True if the second geometry is wholly inside the first geometry.
- Overlaps – True if the intersection of the geometries results in a value of the same dimension as the geometries that is different from both of the geometries.
- BBOX – True if the geometry is within the bounding box envelope provided. This is functionally equivalent to Not Disjoint(....).

A comparison operator is used to form expressions that evaluate the mathematical comparison between two arguments. There are 9 comparison operators defined in the specification, namely:

- PropertyIsEqualTo – True if the two expressions are equivalent. Most implementations convert the expressions into an internal representation and compare them directly. This can lead to some problems if it is not obvious what type a string should be converted to. For example 1.0 will most often be read as a floating point number which means it won't equal "1.0" if it is stored in a database or shapefile as a string.
- PropertyIsNotEqualTo – True if the two expressions are different, the opposite of PropertyIsEqualTo.
- PropertyIsLessThan – True if the first expression is less than the second.
- PropertyIsLessThanOrEqualTo – True if the first expression is less than or equal to the second.

- PropertyIsGreaterThan – True if the first expression is greater than the second expression.
- PropertyIsGreaterThanOrEqualTo – True if the first expression is greater than or equal to the second.
- PropertyIsLike – True if the property of the feature matches the literal expression provided. There are three attributes that must also be provided, namely wildCard which is the character that in the match expression can match as many characters as it can; singleChar which matches any one character; and escapeChar which makes the following character have its literal value instead of its special value. For example:

  if wildCard=* singleChar=. And escapeChar=\ then
  a*b will match aaab azb asddhnddndnnb ab but not nnvnb ammdmm or cat
  a.b will match azb acb but not ab or accb or azcb
  a\.b will only match a.b

- PropertyIsNull – True if the property named is null (or has no value) this is different from 0.
- PropertyIsBetween – True if the property named is between the upper and lower boundaries provided, and both boundaries are inclusive.

There are three logical operations defined in the specification. Specifically:

- And – combines two filters and returns true if they both evaluate to true.
- Or – combines two filters and returns true if either or both evaluate to true.
- Not – returns true if its enclosed filter is false and vice versa.

At the time of writing, GeoTools is in the process of moving to using the GeoAPI interfaces of these filters from the GeoTools interfaces. However, from a user point of view there is no real change in use. User code generally constructs filters and associated expressions using a FilterFactory, which hides the actual implementation of the filter from the end user code.

8.5.1.4 Styling and Rendering

Styling and rendering are the means of converting the abstract representation of a geographic feature into an actual drawing of a map on the screen or on paper. The GeoTools styling system is based on the OGC Styled Layer Description (SLD) language (OGC 2002). The GeoTools styling package allows programs to control the following characteristics:

- colour of lines and fills
- size and type of symbol for points
- text properties (e.g., position, font, colour) for labels, and
- rules for all above based on feature attributes.

In SLD, the StyledLayerDescriptor element is the root of the document. It can contain a NamedLayer with UserStyle containing a FeatureTypeStyle element in it, or it may just contain a list of FeatureTypeStyle objects. This is represented as a

Style object within the GeoTools library, which can contain a collection of FeatureTypeStyle objects.

A FeatureTypeStyle declares a part of a style that is specifically geared toward a FeatureType. That is, features will be rendered according to this FeatureTypeStyle only if their FeatureType is the same as the FeatureType declared in the FeatureTypeStyle or a descendent. A FeatureTypeStyle contains one or more rules. A Rule contains filters that will decide whether features will be displayed or not, specifically:

- A minimum and maximum map scale. If set, and the current map scale is outside the specified range, the rule won't apply and its symbolizers won't be used;
- A Filter that is applied to the features. Only the features matching the filter will be painted according to the Rule symbolizers;
- As an alternative, the rule can have an "else filter". This special kind of filter catches all of the features that still haven't been symbolized by previous rules with a regular filter.

Within the library a Rule object is used to represent the SLD rule element and to hold references to a filter object as defined above. Finally, a Symbolizer describes how to represent a feature on the screen based on the feature's contents (geometry and attributes). Each rule can have one or more Symbolizer attached to it. There are five symbolizers that can be used in a rule depending on the type of feature to be drawn. They are:

- Line Symbolizer – This symbolizer is used to draw lines, such as roads or rivers. It contains a single stroke element, which most commonly takes a stroke parameter to set the colour and a stroke-width parameter to set the width of the line in pixels. In the next version of the specification users will have the option of specifying the width in map units (e.g. feet, metres).
- Polygon Symbolizer – A PolygonSymbolizer is used to render a polygon. It takes a stroke and a fill element. The stroke is exactly the same as for a line. The fill element commonly takes a fill parameter to set the colour of the polygon.
- Point Symbolizer – A PointSymbolizer allows a user to render a point feature on a map such as a building. A Graphic element contains the mark, which is identified by its WellKnownName. A graphic element may contain a list of marks and external graphics in order of preference so that the server can work through the list from first to last and select the first object it can draw.
- Text Symbolizer – A textSymbolizer is used to render text onto the map to label a feature. It takes a Label element, which is what the label says, and a font element to define which font to use.
- Raster Symbolizer – GeoTools currently has no raster symbolization support, as it is only in the latest version that the display of raster data has been added.

To draw a map, a user builds a MapContext object, which consists of a series of map layers and a renderer. Each MapLayer object contains a reference to a datasource and a style. When the map context is rendered, each layer is passed to the renderer. The renderer iterates through the datasource, and applies the rules defined in the style to determine if each feature should be drawn. If the feature is to be rendered, the renderer applies the symbolization defined by the symbolizer to the geometry of the feature and paints it to the graphics object supplied to the renderer. This is repeated for each rule in the style for each layer in the map.

To optimize this potentially lengthy process, the renderer stores the Java shape representation of each feature after it has been converted for painting. The GeoTools renderer also carries out several other optimization steps to improve rendering performance, such as automatic decimation of features where the number of points in the geometry would give greater accuracy than could be represented by the screen resolution.

8.5.1.5 Spatial Referencing

The GeoTools spatial referencing system is almost entirely the work of Martin Desruisseaux and comes originally from the SEAGIS system (http://seagis.sf.net), though there has been much work undertaken to improve and extend the original system. The referencing system is designed to be very flexible and to support almost any type of geographic projection possible. There is a cost to this of course, and it is the production of a very complex system which can be quite intimidating to the novice developer. To help with this, GeoTools provides a collection of helper classes that provide predefined projections as defined by the European Petroleum Spatial Group (EPSG).

By including one of the EPSG database plugins in an application's classpath, it becomes possible to use the CRS utility class to decode strings of the form "EPSG:4326" into CoordinateReferenceSystem objects. These objects can be used to create transform objects, which are used to convert geometry objects from one projection to another. The CRS utility class also provides methods to query projections for their valid geographic area, ellipsoid, and so on.

8.5.2 Extensions

GeoTools provides several extension packages that build on the library to provide additional functionality. These modules are not essential to the working of the library, but provide useful functions for various applications.

8.5.2.1 Graphs

This package provides the ability to convert a collection of features into a graph structure. This allows users to carry out topological operations on the graph. For example, once a set of features has been expressed as a graph, it is possible to calculate a Delaunay triangulation easily and then re-express this as a feature collection that can be used in the rest of the GeoTools library.

The graph module is also used to implement a validation module that allows programs to check the topologic correctness of geometries that are created. This allows programs such as GeoServer to carry out basic checks on input data before they are added to a database.

Finally, the graph capabilities of the GeoTools library can be used to construct routing applications using its implementation of Dijkstra's shortest path algorithm (Dijkstra 1959). This method can calculate the shortest path (or least cost path) from a selected node in a graph to every other node in the graph.

8.5.2.2 ColorBrewer

ColorBrewer is an online tool designed to help users select good colour schemes for maps and other graphics (Brewer et al. 2003). ColorBrewer includes 35 basic schemes with different numbers of classes for over 250 possible versions. Each scheme has CMYK, RGB, Hex, Lab, and AV3 (HSV) specs for the colours. The basic ColorBrewer software is designed simply to list colour specs for useful schemes so the user can apply them in the mapping software.

The GeoTools library takes these schemes and encodes them as style objects the main library can use for styling and rendering operations. This allows users to produce chloropleth and other thematic maps with well designed colour schemes. The ColorBrewer tool also gives users hints as to which colour schemes are good for different media, such as display on a computer screen or photocopying, or that are best for viewing by colourblind users.

8.6 Future Work

GeoTools is a living project which is constantly in development with currently over twenty active developers. These developers are a mix of employees of organizations that make use of the GeoTools library in their own applications, such as Refractions Research (developers of uDig), The Open Planning Project (developers of GeoServer), and the Pennsylvania State University (developers of GeoVISTA *Studio* – see Chap. 10). Other developers work for consultancies that provide commercial support to other organizations using these and other OS tools. Still other developers are students at universities around the world. Recently this group has been increased by the support of Google's Summer of Code program, which pays for summer internships to open source projects (http://code.google.com/soc/2008/).

As a result of this diverse development team and the distributed leadership approach, future work in GeoTools may seem at first glance somewhat random and uncoordinated. In general, something will get implemented if enough people want it done badly enough, or if a developer or user sources funding to pay for the work. For example, if a dependant project requires a specific feature to be developed, the developers of that project can either build that functionality inside their own project, or if they feel it would be generally useful they can contribute that code to the GeoTools library. If they contribute it to the library, then there is a greater chance that others will make use of the code. This helps to enhance and further develop the library.

The overall aim of the PMC is to improve the reputation of GeoTools for stability by more strictly enforcing the policy of deprecating methods for a whole release cycle before deleting them, and by making releases more regular. This has become more important as more projects become dependent on the library. It was no longer acceptable for a group of determined developers to redesign the feature model over one weekend, no matter what the performance gains, if it meant that all dependant projects ceased to work as a result of these efforts. By introducing continuous development builds and rigorous unit testing requirements for new code, the GeoTools main build works directly from the repository much more often, even if some people still chafe under the restrictions to their coding freedom.

At the time of writing, GeoTools is in the process of changing its feature model to allow the support of more complex features than are currently allowed. However, this time the changes are being done in a controlled manner to avoid breaking too much, rather than the "big bang" approach of previous years. The pace of change in the library has also slowed from an end user perspective by the adoption of the GeoAPI interface library which, as it is overseen by the OGC, changes more slowly than many open source projects.

GeoTools has grown over the last ten years from a small individual research project to become a multinational project that is actively developed by more than 20 developers. The library is used by three large open source projects and numerous smaller ones. It is also used as the basis of successful consultancies by several developers who make a living selling services using software developed with GeoTools.

GeoTools also provides an excellent introduction to any student seeking to gain a greater understanding of how GIS operations work in theory and practice without being restricted by copyright and software patents. An interested student can write a simple program to implement, for example, buffering of a feature and be able to see how it is actually implemented in the underlying code. More advanced students can experiment with implementing different methods from theory to see the practical advantages and disadvantages of the different methods.

As GIS becomes more technical, developers and users in general have two distinct choices in their use of the available tools. On the one hand, they can accept the standardized functions of their commercial software packages and simply learn which button to push for any operation, or they can choose solutions which allow them to look "beneath the hood" to see the workings of the software so they can truly understand what is going on when they press a button. Only the latter path provides a true education that will help advance careers, and only by choosing an OS approach will freedom be provided to explore the workings of the software that is available.

References

Brewer CA, Hatchard GW, Harrower MA (2003) ColorBrewer in print: A catalog of color schemes for maps, Cartography and Geographic Information Science 30(1): 5–32

Dijkstra EW (1959), A note on two problems in connexion with graphs. Numerische Mathematik. 1 S. 269–271

Kingston R, Carver S, Evans A, Turton I (2000) Web-based public participation geographical information systems: An aid to local environmental decision-making. Comput Environ Urban Syst 24(2): 109–125

Macgill J (2001) Using flocks to drive a geographical analysis engine, PhD Thesis, School of Geography, University of Leeds, UK

Macgill J, Openshaw S (1998) The use of flocks to drive a geographic analysis machine, Proceedings of GeoComputation'98. http://www.geocomputation.org/1998/24/gc24_01.htm

Open Geospatial Consortium (2002) Styled layer descriptor (SLD) implementation specification, document 02-070, http://portal.opengeospatial.org/files/?artifact_id=1188

Open Geospatial Consortium (2004) Filter encoding implementation specification, document 04-095 http://www.opengeospatial.org/standards/filter

Openshaw S, Turton I, Macgill J, Davy J (1999) Putting the geographical analysis machine on the internet. In Gittings B (ed) Innovations in GIS. Taylor and Francis, London

Chapter 9
GRASS GIS

M. Neteler, D.E. Beaudette, P. Cavallini, L. Lami and J. Cepicky

Abstract GRASS is a full featured, general purpose Open Source geographic information system (GIS) with raster, vector and image processing capabilities. There has been constant development of the software since 1982, with recent major improvements reflecting renewed efforts by the international development team to make it one of the core components of the Open Source geospatial software stack. It can handle 2D and 3D raster data, includes a topological 2D/3D vector engine, network analysis functions, and SQL-based attribute management. This chapter presents an overview and practical examples of the GRASS 6 capabilities relevant to environmental and planning applications including new functionality. Enhancements to 3D visualization and approaches to environmental models are also discussed, as well as image processing routines pertaining to LIDAR and multi-band imagery. Integration of GRASS with other Open Source software packages for geostatistical analysis, cartographic output and Web GIS applications are described. Trends for future development are also discussed.

M. Neteler
Fondazione Mach - Centre for Alpine Ecology, 38100 Trento (TN), Italy,
e-mail: neteler@osgeo.org

D.E. Beaudette
Department of Land, Air and Water Resources, University of California, Davis, CA 95616, USA,
e-mail: debeaudette@ucdavis.edu

P. Cavallini
Faunalia, Piazza Garibaldi 5, 56025 Pontedera (PI), Italy, e-mail: cavallini@faunalia.it

L. Lami
Faunalia, Piazza Garibaldi 5, 56025 Pontedera (PI), Italy, e-mail: lami@faunalia.it

J. Cepicky
Help Service - Remote Sensing s.r.o., Cernoleska 1600, 25601 - Benesov, Czech Republic,
e-mail: jachym.cepicky@gmail.com

G.B. Hall, M.G. Leahy (eds.), *Open Source Approaches in Spatial Data Handling*.
Advances in Geographic Information Science 2,

9.1 Introduction

The Geographic Resources Analysis Support System (GRASS, http://grass.osgeo.org) was born in the early 1980s at the Construction Engineering Research Laboratory (CERL) of the United States Army as a software management tool for military applications. It has evolved into one of the most comprehensive, general purpose free and open source for geospatial (FOSS4G) systems in existence. GRASS is a raster/vector geographic information system (GIS) combined with integrated image processing and data visualization subsystems. It includes hundreds of modules for management, processing, analysis and visualization of geo-referenced, spatial data. It includes unique algorithms and methods that are implemented in a portable environment making it one of the few GIS in the world that works under a variety of different operating systems and platforms. GRASS is fundamentally a desktop GIS, however recent efforts are in place to extend functionality to Web-based and network-programmable interfaces.

The limitations of computer hardware in the early 1980s, coupled with the innovation of skilled and motivated young researchers lead to a first working version of GRASS. It was designed as a modular system written in the C programming language. The advent of the Internet facilitated the spread of the software among universities and other governmental organizations. As CERL and GRASS evolved through the late 1980s and early 1990s, CERL created the Open GRASS Foundation which evolved into the Open GIS Consortium (OGC, now known as the Open Geospatial Consortium). The discontinuation of development of GRASS by CERL in 1996 lead to the formation of the GRASS Development Team two years later.

A first release as Free Software under the GNU General Public License (GPL) was published in October 1999. Based on academic efforts, GRASS is today developed by a worldwide group of scientists, programmers, power users and enthusiasts. The project is attractive to a new generation of users and developers due to the modernization of the software. The release of GRASS 6 reflects new efforts to one of the core components in the FOSS4G software stack. Quality management is enforced by transparent development methods such as a centralized source code repository (in a concurrent versions system (CVS) server) and the immediate peer-review of source code changes by email broadcast.

In 2006, the Open Source Geospatial Foundation (OSGeo, http://osgeo.org) was created to bundle several FOSS4G software projects, including GRASS, into a joint organization. This chapter presents an overview of the GRASS 6 capabilities relevant to environmental and land management applications, including the updated and new tools for 2D and 3D raster data, the vector modules based on the redesigned topological 2D/3D vector engine, and SQL-based attribute management. Approaches to linking with other FOSS4G tools and environmental models are also discussed.

Table 9.1 Functionality classes of GRASS commands

Prefix	Function Class	Type of Command
d.*	display	graphical output
db.*	database	database management
g.*	general	general file operations
i.*	imagery	image processing
m.*	misc	miscellaneous commands
ps.*	postscript	map creation in Postscript format
r.*	raster	2D raster data processing
r3.*	3D raster	3D raster data processing
v.*	vector	2D and 3D vector data processing
d.*	display	graphical output
db.*	database	database management
g.*	general	general file operations
i.*	imagery	image processing
m.*	misc	miscellaneous commands

9.2 The Structure of GRASS

GRASS GIS is a suite of modules designed to facilitate rapid analysis of raster, vector and tabular data. These modules were designed to be as robust and efficient as possible at performing a specific task. Following the UNIX methodology of chaining together multiple small programs to perform complex tasks, spatial analysis is usually performed in GRASS by linking together multiple modules either directly through pipes (an operating system construct, used to pass data between processes), or indirectly through intermediate files. Each of the modules expects a small number of command line arguments, and either returns a new spatial data file (raster, vector, etc.), or modifies an existing map or attribute table in-place. Commands are clearly organized by functionality through a command class prefix shown in Table 9.1.

Although this approach gives tremendous power to a user who learns to wield the full set of GRASS modules, it does not lend itself well towards the graphical user interface-driven desktop application that most people are familiar with. Indeed, the core GRASS code is better described as a professional analytical engine rather than a desktop GIS. However, several graphical interfaces to the command line-driven modules have been produced over the years with the TclTk toolkit, and more recently with the wxWidgets Python toolkit (http://www.wxwidgets.org). GRASS contains several libraries dedicated to simplifying access to and manipulation of very large files including, the "rowio" library, "segments"/"spatial index" libraries (GRASS raster and vector data), and interoperability (reading and writing external formats such as GeoTIFF). A parallelized numerical library has recently been added to the GRASS code base, which assists with converting complex single-thread algorithms into multi-threaded versions that can fully take advantage of multi-processor or multi-core hardware.

9.2.1 Installation

GRASS software can be downloaded freely from the main GRASS Web site (http://grass.osgeo.org). This site is mirrored in several countries for faster access, including the United States GRASS mirror at http://grass.ibiblio.org. Source code (the portable version for all operating systems) as well as the latest ready-to-install binaries for GNU/Linux, MacOSX and MS-Windows (native or optionally with the Cygwin tools) are all available on this site. GRASS is also available on CDROM/DVD from various providers. The MS-Windows version of the popular Quantum GIS (QGIS, http://qgis.org) package includes a native GRASS installation. QGIS is a user friendly geographic data viewer with some analytical capabilities that runs on GNU/Linux, Unix, MacOSX, and MS-Windows. It supports vector, raster, database formats, OGC Web Services and includes a GRASS toolbox. QGIS is licensed under the GNU GPL.

9.2.2 Functionality and Command Structure

GRASS contains over 300 programs and tools to render maps and images both on-screen and on paper, to manipulate raster and vector data, to process multi spectral and time series image data, and to create, manage, and store spatial data. As noted earlier, GRASS uses both an intuitive graphical user interface as well as command line syntax for ease of operations. Approximately 200 of the modules that are available in GRASS GIS are integrated in pull down menus. The most frequently used modules can thus be easily accessed with the use of a mouse.

GRASS 6 introduces a new topological 2D/3D vector engine and support for vector network analysis. Attributes are managed in DBF files or a SQL-based database management system (DBMS). The NVIZ visualization tool displays raster data, 2D and 3D vector data as well as voxel volumes. GRASS project databases ("locations") can be auto-generated by a European Petroleum Survey Group (EPSG) code number or from geo-coded data sets. GRASS is fully integrated with GDAL/OGR libraries to support an extensive range of raster and vector data formats (see Chap. 5 of this book), including the OGC Simple Features specification. To facilitate usage by non-English speakers, user messages have been translated to various languages including Asian languages. GRASS supports work groups through its *LOCATION/MAPSET* directory structure concept which can be set up on shared/network disk devices. Through this, team members can simultaneously work in the same project's database.

9.3 GRASS Features

Although GRASS has evolved into a general purpose GIS, its major strength remains environmental modeling and analysis. With respect to older versions, significant improvements have been implemented to strengthen its viability. In this regard, interoperability is seen as a crucial element in GRASS development, leveraging from widely used data exchange libraries. Key motivations for using the software are its strong analytical capabilities for both raster and vector data (in 2D and 3D space) as well as graphical data exploration and visualization.

9.3.1 Data Exchange: Interoperability

GRASS data are maintained in their own directory structure in so-called "locations" and "mapsets". The idea behind this structure is to provide a multi-user environment with access control. Locations can be maintained on a centralized server. Extensive capabilities of data exchange are essential for daily GIS work. For interoperability, GRASS profits from an external project, the GDAL/OGR library, which allows the conversion among many raster and vector formats, including the internal GRASS formats. This library is also used by global data vendors as well as in some proprietary GIS applications. Additional GRASS modules allow importing from and exporting to other formats.

GRASS uses a topological vector architecture. To support Simple Features vector data (conformal to OGC standards), these data sets are converted upon data import. Conversely, export into Simple Features is also possible. Instead of importing data sets, vector maps may also by linked to the GRASS database as virtual maps. The module for importing vector maps, *v.in.ogr*, uses the OGR library (for a full list of supported formats, see the GDAL/OGR Web site, http://www.gdal.org/). This module also allows merging of a series of different vectors, limiting the import to a user defined spatial subset, and creating a new GRASS location on the basis of the projection system of the imported data. Additional modules allow the management of a variety of data formats, including ASCII files, 2D and 3D drawings from computer assisted design (CAD) software, common vector formats, Gazetteer and global positioning system (GPS) data, as well as data import from Web Feature Services (WFS).

The *r.in.gdal* module, which uses the GDAL library, enables raster datasets to be imported. The module also allows the import of single bands from multi-band imagery, and the creation of a new GRASS location from the imported dataset. Additional modules allow the import of various ASCII and binary files including all common raster and imagery formats including elevation models, aerial and satellite imagery, and data from Web Mapping Services (WMS). It is also possible to create 3D raster maps from 2D map slices, or from importing 3D volume files.

Once in GRASS, raster maps can be converted to vector maps and vice versa. Vector geometries can also be converted between different types (points, lines, polygons), and a series of raster layers can be transformed into raster volumes and vice versa. 3D points can also be converted or interpolated to raster volumes. Exporting files is essentially similar, with the same formats supported (with minor differences). GRASS raster and vector maps can be exported also to rendering software (e.g., POV-Ray or Paraview). 3D fly-throughs can be created as MPEG format animations.

9.3.2 Raster Data Analysis: Pixels and Voxels

Raster analysis is the historical core of the GRASS project. In addition to 2D raster maps (pixel-based), GRASS can also manage 3D (volume) raster maps (voxel-based). Many modules are available for raster processing, and they can be subdivided into the following groups:

- *Map management*: editing, extent, management of null values, resampling, re-projecting etc.;
- *Colour management*: setting colour tables, weighted merging and splitting of RGB/HIS and normal maps;
- *Buffer creation*: single and multiple buffers;
- *Mask creation*: to apply analyses to a limited subset of the working map;
- *Proximity analysis*: analysis through a moving window to create maps in which every cell is the result of a function applied to the surrounding cells; calculation of minimum distances among raster objects;
- *Map overlay*: statistics, temporal series, patching maps, etc.;
- *Solar radiance and shadowing*: calculation of solar irradiation and shadows on the basis of date and time;
- *Terrain analysis*: calculation of total cumulative cost of movement over a digital terrain model (DTM) or a cost map, search of a minimum cost path, profile analysis, calculation of exposure and slope maps, texture analysis, visibility analysis;
- *Raster transformation*: finding islands of contiguous cells of the same type, growing and linearizing raster islands;
- *Specialized models*: hydrology, landscape ecology, fire spread;
- *Reclassification*: on the basis of user rules, of island size, recoding and rescale;
- *Random raster maps*: generating raster points (also on the basis of a second raster map);
- *Surface generation*: density maps (kernels) from vector points, planes, fractal surfaces, Gaussian derivates, random surfaces with spatial dependency;
- *Surface interpolation*: bilinear, bicubic, inverse distance weighted (IDW), and regularized splines with tension (RST);
- *Reports and statistics*: general statistics, correlation and linear regression between raster maps.

GRASS raster map processing is always performed in the current region settings. That is, the current region extent and current raster resolution is used. If the resolution differs from that of the input raster map(s), on-the-fly re-sampling is performed (nearest neighbour re-sampling). If this is not desired, the input map(s) has/have to be re-sampled beforehand with one of the dedicated modules. GRASS applies two general rules in this context:

1. Raster/imagery output maps have their bounds and resolution equal to those of the current region.
2. Raster/imagery input maps are automatically cropped/padded and rescaled (using nearest-neighbour resampling) to match the current region.

9.3.3 Vector Data Analysis

The vector engine of GRASS has been completely overhauled during the last few years. Geometry and attribute management are clearly separated, giving more flexibility in data storage. The vector geometry was extended to manage 2D and 3D vector data. Vectors are fully topological, with spatial relationships (e.g., connection, adjacency, inclusion) embedded in the information linked to each vector element. For instance, a border line between two polygons is not replicated, but is singular, and linked to the left- and right-sided polygon centroids. This new internal vector data format is also portable across 32bit and 64bit platforms.

A new spatial index system as well as category index accelerates data access. The support of topology enables the user to perform data cleanup to ensure data consistency. Support for vector map overlays, intersections and extraction of features is implemented. A new digitizing tool permits users to create or update vector features with attributes on-screen. The attribute management includes full and flexible integration of database management systems (PostgreSQL, MySQL, DBF, SQLite and ODBC are currently supported). Structured query language (SQL) statements, used to manage attributes, are directly passed to the underlying database system. The set of supported SQL commands depends on the RDMBS and selected driver. Graphical updating of vector attributes has been implemented as well. Available vector data types are:

- *Point*: a single data location;
- *Line*: a directed sequence of connected vertices with two endpoints called nodes;
- *Boundary*: the border line to describe an area;
- *Centroid*: a point within a closed boundary;
- *Area*: the topological composition of centroid and boundary;
- *Face*: a 3D area;
- *Kernel*: a 3D centroid in a volume (*in development*);
- *Volume*: a 3D corpus, the topological composition of faces and kernel (*in development*).

GRASS vector maps may include different vector types (points, lines, areas etc.) in the same layer. Available vector analysis methods are quite advanced, and are categorized into several groups:

- *Map management*: editing, topology rebuilding, cleaning up of non-topological imported vectors, adding centroids to polygons, merging and splitting of elements, geometry conversions (2D and 3D; rebuilt on the basis of a DTM; extrusion), re-projection;
- *Attribute management*: connecting vectors to attribute tables (on various database backends);
- *Reports and statistics*: general statistics, export of geometric characteristics to database, normality tests on point distribution, correlation of values and corresponding raster cells;

- *Extraction and selection*: on the basis of a geometry or a table query, selection of a vector on the basis of the relations with the geometries of another vector buffer;
- *Proximity analysis*: calculation of minimum distance matrix among objects;
- *Map overlay*: patching and overlaying vectors, management of attributes;
- *Generation of reference maps*: creation of grids of polygons or points;
- *Reclassification*: management and reclassification of categories associated to geometries;
- *Points*: import of point vectors from an x,y[,z] table, creation of random points, minimum convex polygons, Voronoi tessellation and Delaunay triangulation, raster values extraction;
- *Lidar data analysis*: outlier detection, Digital Surface model (DSM) and DTM separation and interpolation;
- *Network analysis*: creation of a network, finding the shortest path, allocation of sub-nets for nearest centers, Steiner tree, or resolution of the traveling salesman problem;
- *Linear referencing system*: creation, editing and management.

9.3.4 Attribute Data Storage via a DBMS

GRASS provides tools for management and analysis of vector maps, including the attributes that can be stored in a DBMS. GRASS can be connected to various relational DBMS (RDBMS) and embedded databases. It supports SQL to create, retrieve, update, and delete data from RDBMS. GRASS unifies the different drivers in an abstraction layer named the database management interface (DBMI) to assist the user.

Usually an attribute table is linked to vector geometry data. The attribute table must contain a column named "cat" to store the category numbers (the vector IDs) which connect the individual vector objects to attributes. Each table row corresponds to a category number. Several vector objects can be assigned to the same category numbers (table row).

It is possible to link the geographic objects in a vector map to one or more tables. Each link to a distinct attribute table is called a layer. A link defines the database driver, the database name, and the table name that are used. Each category number in a geometry file corresponds to a row in the attribute table. Using *v.db.connect*, layers can be listed or maintained. GRASS layers do not contain any geographic objects, but they consist of links to attribute tables in which vector objects can have zero, one, or more categories. If a vector object has zero categories in a layer, then it does not appear in that layer. Some vector objects may appear in some layers but not in others. The practical benefit of this approach is that it allows placement of thematically distinct but topologically related objects into a single map (e.g., forests and lakes). These virtual layers are also useful for linking time series attribute data to a series of locations that do not change over time. By default the first layer is

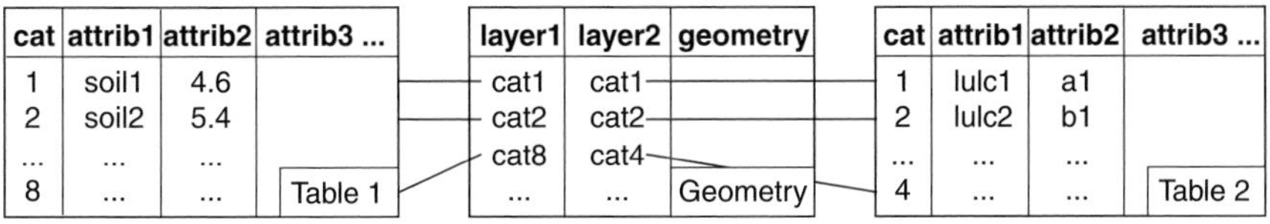

cat	attrib1	attrib2	attrib3 ...
1	soil1	4.6	
2	soil2	5.4	
...	...	...	
8	...	...	Table 1

layer1	layer2	geometry
cat1	cat1	
cat2	cat2	
cat8	cat4	
...	...	Geometry

cat	attrib1	attrib2	attrib3 ...
1	lulc1	a1	
2	lulc2	b1	
...	...	...	
4	...	...	Table 2

Fig. 9.1 Example for layer concept in GRASS vector files: geometry file with two layers connected to two different attribute tables (linking field "cat")

active (i.e., the first table corresponds to the first layer). Further tables are linked to subsequent layers. Figure 9.1 illustrates the layer concept with an example for a vector map with two layers, connecting two different attribute tables to the vector geometry.

9.3.5 Visualization: 2D, 3D and Animations

Besides the creation of 2D maps, GRASS also provides the possibility to create impressive 3D visualizations and animations of surfaces and volumes using the NVIZ module (Fig. 9.2). This module uses raster data to define both height and attribute data, and to allow the overlay of vector data (points, lines and areas). It also has illumination tools to add shadow effects. After starting NVIZ, a graphic window and a control window are opened, and the individual features are displayed. A series of images can be saved and combined into a GIF animation or an MPEG movie file via external programs. This is especially interesting when analyzing changes over time.

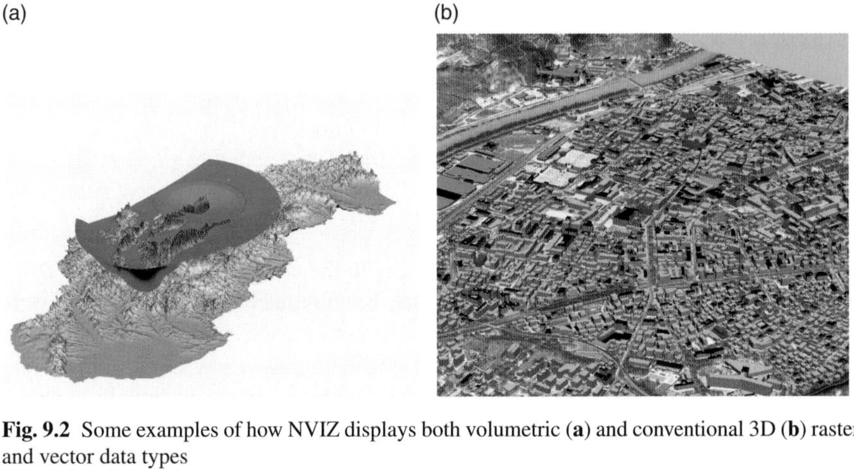

(a) (b)

Fig. 9.2 Some examples of how NVIZ displays both volumetric (**a**) and conventional 3D (**b**) raster and vector data types

9.4 Raster Applications

Given the raster data processing origins of GRASS, it is not surprising that the raster modules are among the oldest and best developed in its toolbox. Several common applications of the GRASS raster modules such as surface analysis, geomorphometric classification, interpolation, and solar radiation modeling are included in this section. A more complete discussion of these modules can be found in Neteler and Mitasova (2008).

9.4.1 Cost Surface Analysis: Wilderness Navigation

Cost surface analysis is a method of automatically identifying a path which minimizes the amount of accumulated "cost" between a series of points. Conceptually, this approach is very similar to the actions of a mountain hiker seeking to minimize the total change in slope encountered while traversing an area. It follows the principle that walking *up or down* steep hills is hard, and walking a *long distance* up or down a steep slope is even harder. In GRASS, the modules *r.cost* and *r.walk* are used to compute accumulated cost from some start point to an end point. The module *r.drain* is then used to identify the path of least cost between end and start points, akin to the movement of a drop of water down an elevation gradient. A slope map is commonly used as an input to *r.cost*. However, more complex systems can be modeled by altering a slope map with *r.mapcalc*.

Consider the example of a trip through rugged terrain between numerous subalpine lakes. The region is mostly wilderness, and therefore trails are few and do not directly connect specific lakes that fall along a desired route. Identifying the paths between points of interest which minimize slope traversal has been done with paper topographic maps for decades. Cost surface analysis is well suited to this type of problem. For example, the task of navigating between Bill Lake and Bonnie Lake shown in Fig. 9.3(a) is a good case in point. In this example, additional sources of "cost" are added to the slope map as follows:

- traversing lakes is not acceptable (extreme travel cost added to slope map);
- traversing wooded areas is faster due to shading (moderate travel cost subtracted from slope map).

By specifying start and stop points, along with the composite travel "cost" map, the least-cost path between these points can be computed via the GRASS commands *r.cost* and *r.drain*. It is possible to see how *r.drain* actually simulates the flow of a drop of water down an elevation surface in an energy minimizing operation (Fig. 9.3(b)). However, in this case, the elevation surface is actually a travel cost map (computed by *r.cost*). The path of the theoretical drop of water represents the least-cost route from the starting point to the end point. Note that the start point for *r.cost* (point of origin) is actually the end point for *r.drain*. By specifying very large travel costs for traversing lakes (peaks), and slightly lower travel costs for wooded areas, the computed least cost path goes "around" lakes and "through" wooden areas.

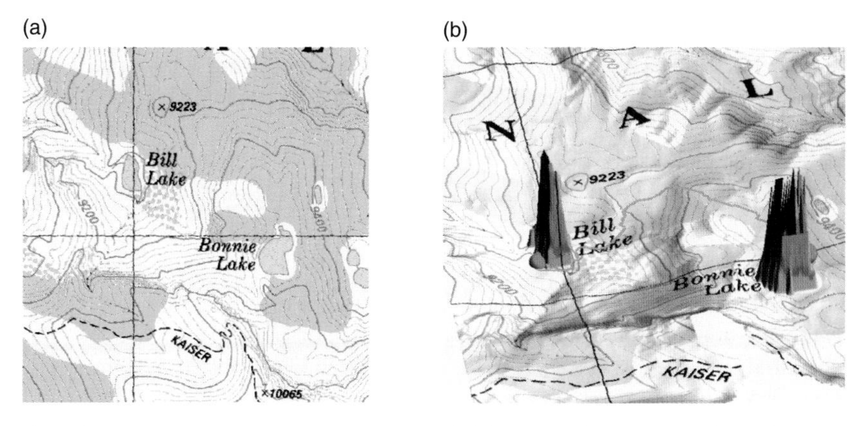

Fig. 9.3 An example least-cost path calculation. A hiker wishes to travel from Bill Lake to Bonnie Lake (**a**); the accumulated cost surface and least-cost path between the two points (**b**)

9.4.2 Interpolation Functions

GIS are often used to interpolate continuous raster surfaces from regularly, irregularly, or even scarcely distributed point data. These surfaces usually represent elevations, temperatures, or other continuous phenomena. Interpolations can also be performed from contour data. Re-sampling to a different resolution can also be seen as a special case of interpolation.

GRASS supports a range of re-sampling and interpolation methods such as nearest neighbour re-sampling, bilinear and bicubic interpolation (*r.resamp.interp*), inverse distance weighted (IDW in *v.surf.idw*, Lo and Yeung 2006; Issaks and Srivastava 1989) and regularized splines with tension (RST) in *v.surf.rst* as well as *v.vol.rst* for volumetric data (Mitasova and Mitas 1993; Mitasova and Hofierka 1993).

Data interpolation in GRASS is explained in detail in Neteler and Mitasova (2008). The RST method especially gives much control over the behaviour of the interpolator through the smoothness and tension parameters. To obtain reasonable results, an integrated cross-validation algorithm helps to find input parameters which minimize the interpolation error. Integrated segmentation allows interpolation of massive data sets. Additional related methods are re-sampling of point data and raster maps using spatial aggregation (*r.in.xyz* and *r.resamp.stats*) and temporal aggregation (*r.series*).

9.4.3 Geomorphological Analysis

Landscape processes, including water and sediment flow, are both caused and influenced by the geometry and properties of the land surface. Basic land features (e.g.

planes, pits, peaks, ridges etc.) can be identified with the module *r.param.scale*. GRASS includes also an extensive set of modules for deriving land-surface parameters and performing spatial analysis that involves elevation data (Hofierka et al. 2008).

Fundamental parameters of the surface can be calculated using partial derivatives of the mathematical function describing the surface as local parameters (based on a point and its immediate surroundings):

- Slope (steepest angle of the slope);
- Aspect (slope orientation, direction of gradient, steepest slope or flow direction);
- Profile curvature (surface curvature in the direction of gradient);
- Tangential curvature (surface curvature in the direction of contour tangent);
- Mean curvature (an average of the two principal curvatures).

Additionally, GRASS provides a wide array of tools to analyze water flow and watershed parameters, starting from basic parameters (flow accumulation, upslope contributing area, stream network, watershed (basin) area, flow path length) to complex algorithms for flow routing:

- Single flow direction to eight neighbouring cells moves flow into a single downslope cell (*r.watershed*);
- Single flow to any direction (D-inf or vector-grid approach) (*r.flow*);
- Multiple flow direction (MFD) to two or more down slope directions (*r.terraflow*, *r.topmodel*);
- 2D water movement simulation based on overland flow differential equations (*r.sim.water*).

Various models and parameters can be combined, and new models can be developed through the use of the map algebra module *r.mapcalc*.

9.5 Vector Applications

The GRASS vector model contains a description of topology. From the user's perspective, topological operations enable verification and enforcement of data integrity and quality. By default, GRASS 6 always builds the topology after importing, creating or modifying a map. A considerable amount of detail about these algorithms can be found in the relevant literature (Lo and Yeung 2006; Neteler and Mitasova 2008)

9.5.1 Overlay Operations and Selections

Geometric operations are performed through specialized modules. By selecting an operator, features from two input maps are geometrically elaborated and the result is written to a new output vector map. A typical set of operators and tools are available:

- *and*: also known as "intersection";
- *or*: also known as "union";
- *not*: features from map A not overlaid by features from map B;
- *xor*: features from either map A or map B but not those from A overlaid by B;
- point-in-polygon selection;
- vector extraction (spatially, or based on SQL statements).

9.5.2 Network Analysis and Linear Reference System

The integrated Directed Graph Library (DGL) provides support for vector network analysis. Available algorithms include shortest path, traveling salesman (round trip), allocation of sources (sub-networks), Minimum Steiner Trees (star-like connections), and iso-distances (from centres). Costs may be assigned both to nodes and arcs. Both directions of a vector line can be used, which permits definition of a forward and a backward direction, and storing of their attributes in the related attribute table. An example for shortest path routing on a vector network is shown in Fig. 9.4.

The following modules are available:

- *Shortest-Path Analysis*: the commands *d.path* and *v.net.path* allow calculation of the least expensive (by default the shortest path as the length of the vectors is used as the measure of cost) distance between two chosen points with two methods. Additional information about the vectors (e.g., speed limit on the road or road status) can be used for calculating a path. Cost information can also be assigned differentially to both vector directions. Attributes of the nodes (e.g., cycle times of the traffic lights at a crossroad) can also be considered.
- *Subnets within a vector network*: a given vector network can be subdivided into sub-networks with the module *v.net.alloc*. For instance, regions in a city can be identified that are best served by a limited number of fire stations.

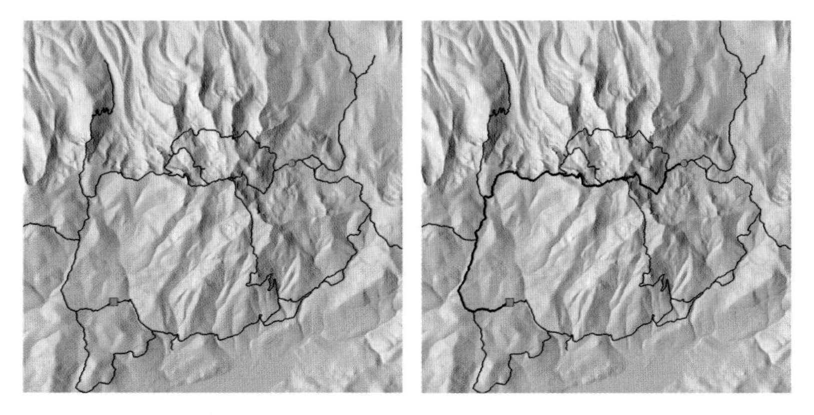

Fig. 9.4 Simple example of a shortest-path calculation from node 1 (*green square*) to node 2 (*red circle*) along a network

- *Minimum Steiner Tree problem*: the optimal connection of nodes within a network (star) can be described by the Minimum Steiner Tree. For instance, several hospitals distributed in a region need new fibre-optical network cables for telemedical services. The aim is to lay the necessary cable along existing roads, while using the least amount of cable to connect all hospitals. The GRASS module *v.net.steiner* is used for this purpose.
- *Traveling-Salesman problem*: the classic problem of calculating the best route joining a series of points (either by distance or time) can be solved with the GRASS module *v.net.salesman*. As an example, a route minimizing cost for a company representative visiting a series of customers can be calculated.
- *Cost analysis*: finding iso-distances (concentric distances around a series of points) on a network can be performed with the GRASS module *v.net.iso*. For instance, "run-length" (e.g. for sewage channel systems) can be calculated based on the vector length or, as in previous modules, other attributes. Another application is the search for reachable points of interest within a five minute walk from a metro-station.

Besides vector networking, GRASS also supports Linear Reference Systems (LRS; Blazek 2005). A LRS is a system where features (points or segments) are localized by a measure along a linear element. The LRS can be used to reference events for any network of linear features, for example roads, railways, rivers, pipelines, electric and telephone lines, water and sewer networks. An event is defined in LRS by a route ID and a segment. A route is a path on the network, usually composed from more features in the input map. Events can be either points or line segments.

9.5.3 LIDAR Data Processing

Airborne Light Detection and Ranging (LIDAR) is one of the most recent technologies in surveying and mapping. A laser on board a plane sends out pulses to the ground in order to determine the distance to an object or surface. This distance is determined by measuring the time delay between transmission of a pulse and detection of the reflected signal. The horizontal and vertical accuracies are in the centimetre range. Up to four range measurements can be performed for each pulse. LIDAR offers various new applications including terrain change analysis to study vertical and horizontal changes in terrain (e.g. beaches) and allowing for calculation of area and volume change. Another application is hazard mapping (e.g. detection of avalanches or other morphological hazards). In the area of energy analysis, solar energy can be estimated at building detail for alternative energy evaluation, since roof inclination and structure details can be derived from LIDAR data. The LIDAR toolset in GRASS provides advanced methods to compute the digital surface (DTM or DSM) based on radial basis functions and spline functions with the Thykhonov regularizer (Brovelli et al. 2004).

The data elaboration procedure is started with the import of LIDAR point clouds (first and last pulse) with *v.in.ascii*. In this case, the topology is not built to avoid redundant computations. Subsequent outlier detection is done with *v.outlier* on both first and last pulse data. At the next step, edges are detected from the last pulse data with *v.lidar.edgedetection*. The DSM (buildings, vegetation etc.) is generated with *v.lidar.growing* from detected edges. The resulting data are post-processed with *v.lidar.correction*. Finally, the DTM and DSM are generated with *v.surf.bspline*.

The *r.in.xyz* module can be used to perform binning of points derived from raw sensor measurements, or coordinates exported from the common American Society for Photogrammetry and Remote Sensing LIDAR Exchange Format (LAS), to grid cells of a given size. This module can use several common statistics such as min, max, mean, and range for cell-wise aggregation. With careful tuning of region settings (extent and resolution), it is possible to generate grid point data where the number of features exceeds the practical limits imposed by the GRASS vector engine. Tests have shown that *r.in.xyz* performs well with over 25 million input features.

9.6 Image Processing

Remote sensing is a rapidly advancing technology for gathering environmental data using a wide range of airborne and satellite platforms, and it plays a major role in spatio-temporal earth surface monitoring. Image data within GRASS are treated identically to raster data. However, several commands are explicitly dedicated to image processing. GRASS supports import of common satellite data and imagery formats, geocoding of imagery data, visualizing (true) colour composites, calculation of vegetation indices, calibration of channels, image classification and image fusion as well as time series processing.

Satellite imagery and orthophotos (aerial photographs) are handled in GRASS as raster maps and specialized tasks are performed using the imagery (i.*) modules. All general operations are handled by the raster modules.

9.6.1 Common Operations

Besides visualization, all common steps of radio and geometric pre- and post-processing are supported. Some of the available techniques are illustrated in the following sub-sections.

9.6.1.1 Visualizing True Colour Composites

The GRASS command *d.rgb* can be used to combine the first three channels quickly to a near natural colour image. The graphical GIS manager (*gis.m*) offers an easy interface to work with colour composites. The procedure assigns each channel to a colour which is then mixed while displayed. With some user optimization of the grey scales of the channels, nearly perfect natural colours can be achieved. The *i.landsat.rgb* can be used to adjust automatically LANDSAT red, green, and blue channels to closely match the colours perceived by the human eye. Channel histograms can be displayed with *d.histogram*. However, this command operates on any raster map.

9.6.1.2 Calculation of Vegetation Indices

A common proxy for chlorophyll content, utilizing both red (LANDSAT channel 3) and near infrared (LANDSAT channel 4) wavelengths, is the normalized difference vegetation index (NDVI; Jensen 2000). A NDVI map is commonly used to infer water stress in plants, vegetative intensity, or potential crop yields. Calculation of this and other vegetation indices can be done in a single step with simple map algebra, as implemented in the GRASS command *r.mapcalc*:

$$\mathrm{ndvi} = 1.0^{*}(\mathrm{tm4} - \mathrm{tm3})/(\mathrm{tm4} - \mathrm{tm3})$$

where:

ndvi the resulting map,
tm3 and tm4 the LANDSAT channels existing as GRASS raster maps.

The command *r.colors* can be then used to create an optimized colour table. The Kauth-Thomas "Tasseled Cap" transformation for LANDSAT-TM sensor data can be performed directly with the GRASS command *i.tasscap*. A comprehensive list of vegetation indices and their respective formulae can be found in Jensen (2000).

9.6.1.3 LANDSAT Operations

The encoded digital numbers of a thermal infrared channel can be transformed to degrees Celsius (or other temperature units) that represent the temperature of the observed land surface. This requires a few algebraic steps with *r.mapcalc* which are outlined in the literature to apply gain and bias values from the image metadata (Neteler and Mitasova 2008).

A downscaling approach commonly applied to channels 1, 2, 3, 4, 5, and 7 (30 meter resolution) can be performed in GRASS with the Brovey transform (*i.fusion.brovey*) method. This approach *fuses* the panchromatic channel (15 metre resolution) with selected channels to produce a new "pan-sharpened" color composite at 15 metre resolution. Colour composites can be displayed with *d.rgb*, as described earlier, or saved with *r.composite*.

9.6.1.4 Time Series Processing

GRASS also offers support for time series processing (*r.series*). Statistics can be derived from a set of co-registered input maps such as multi-temporal satellite data. Common univariate statistics and linear regressions can be calculated. For example, Rizzoli et al. (2007) used MODIS time series data processed in GRASS to predict the highest risk areas for increased tick-borne encephalitis virus activity in the Province of Trento, Italy.

9.6.1.5 Rectification/Ortho-Rectification

GRASS is able to geocode raster and image data of various types, including unreferenced scanned images of maps by defining four corner points, unreferenced satellite data from optical and Radar sensors by defining a certain number of ground control points (*i.group*, *i.target*, *i.points*, *i.rectify*), and orthophotography based on DEM data (*i.ortho.photo*). Also, digital photographs from handheld cameras can be geocoded using a modified procedure for *i.ortho.photo* (Neteler et al. 2005).

Geo-referencing in GRASS is done by first importing the images into a generic x,y (non-projected) location. Several bands of the same images can be treated together, defining a group. The user is then prompted to identify a series of ground control points. A Root Mean Square is automatically calculated during the process, providing a means to keep track of the overall accuracy while inserting points. Referenced images are then transformed into a new geo-referenced location by polynomial transformation.

9.6.2 Image Classification

A thematic map can be generated from one or more input channels. These input channels are usually derived from aerial or satellite data. Multispectral data can be considered as a stack of raster maps with identical spatial references. During the image classification procedure the spectral response of objects is analyzed and assigned to classes. The resulting map contains a set of classes which may represent land use or land cover. An example of unsupervised classification, performed with an IKONOS multi-band image is depicted in Fig. 9.5. Modules *i.class* and *i.maxlik* were used for this analysis.

GRASS supports multiple channels which can be grouped together with the *i.group* module. Then either an automated statistical analysis can be performed on the input channels (i.e., an unsupervised classification), or training areas can be digitized by the user to define known land use/land cover areas (i.e., a supervised classification). GRASS derives spectral signatures for the desired classes and runs the final analysis on all pixels of all input channels, assigning each pixel to a class. In the case of unsupervised classification, the classes are arbitrarily numbered, while

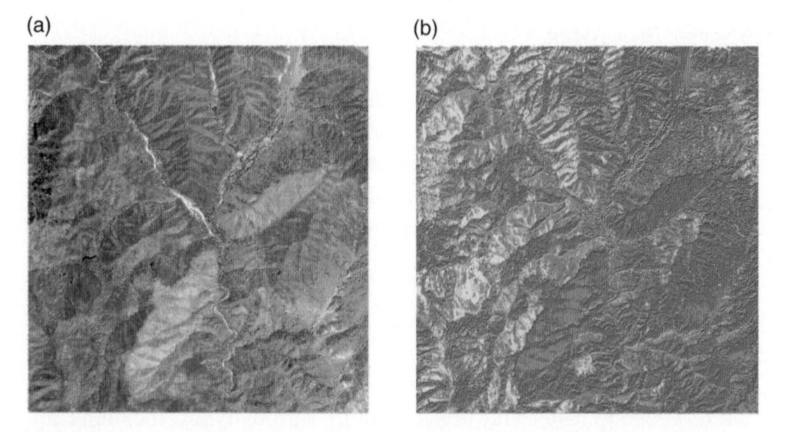

Fig. 9.5 Unsupervised classification of an IKONOS 3-band pseudocolour image (**a**) into 5 discrete classes (**b**)

in the case of supervised classification they correspond to the names of the training areas (Neteler et al. 2005).

Single and multispectral data can also be classified to user-defined land use/land cover classes. In the case of single channel data, segmentation or partitioning of images is used. GRASS supports the following methods:

- Radiometric classification;
- Unsupervised classification (*i.cluster* and *i.maxlik*) using the Maximum Likelihood classification method;
- Supervised classification (*i.gensig* or *i.class*, *i.maxlik*) using the Maximum Likelihood classification method;
- Combined radiometric/geometric (segmentation based) supervised classification (*i.gensigset*, *i.smap*);
- Kappa statistics can be calculated to validate the results with *r.kappa*.

Machine learning, the science of discovering and recognizing patterns in data with algorithms that improve automatically through experience, has been frequently applied to image processing in recent years. Based on samples, a classification or regression function is synthesized in order to predict unknown observations (for example, for a land use/land cover classification). Several machine learning technologies have been implemented in a new set of GRASS modules. The recently added *i.pr* family of image processing modules implement several additional modern classification routines including:

- k-NN (multiclass);
- classification trees (multiclass);
- maximum likelihood (multiclass);
- support vector machines (binary).

Robust estimation of classes is gained with bagging and boosting. Feature normalization within each module allows for the inclusion of predictor variables with non-normal distributions or different ranges. There is planned support for feature selection techniques such as re-sampling and cross-validation, for all classification algorithms available in *i.pr*.

9.7 GRASS Development

The FOSS4G concept offers a license scheme which defines the extent of the usage, modification, and redistribution of original and derived software. For GIS, unlimited access to the source code is of particular interest, as the underlying algorithms are often complex and have significant influence on the results of spatial analysis and modeling. While an average user may not be able to trace errors within complex source code, there are a number of specialists willing to test, analyze and fix any errors identified within the code. This framework is embedded into an Internet-based development model with a high frequency of new releases. The diversity in background and expertise of the developers contributes greatly to network synergy. Overall, this type of software development model leads to faster and more efficient production, along with stable and robust products.

9.7.1 The Development Model: Community-Based

Access to others who are using GRASS, and to the individuals who are actively maintaining the GRASS code base is critical to the OS development model. Several GRASS-related mailing lists and Internet Relay Chat (IRC) channels serve as windows into the GRASS user base and development communities. Additionally, a bug and wish tracking system is used. New and seasoned users alike can find answers, submit bugs, or even contribute code and documentation suggestions through these resources. This style of development fosters a two-way relationship between users and developers. In many cases users find that over the course of a couple of years they can progress from a simple interested party to a part-time contributor of documentation and source code suggestions. Several of the core GRASS development team have followed this path, and they are now in charge of a large and complex code base.

The development tool set includes a server-based code repository with immediate change notification via email for peer-review of the changes. The GRASS Project Steering Committee (PSC) is responsible for granting write access to new developers. To control and improve the source code quality, an automated code monitoring system was established to apply software engineering metrics and clone detection (i.e., identification of undesired source code copies which are identical or rather identical) on every change happening in the source code repository (Bouktif et al. 2006). This GRASS quality assurance system will be further improved and extended in the future and may potentially be made available to further projects under the auspices of the Open Source Geospatial Foundation (OSGeo) (see Chaps. 1 and 2, among others, of this book).

9.7.2 GRASS Programming

GRASS provides a unique opportunity to improve and extend basic GIS capabilities through new code development and the support of the GRASS developer community. The complete GRASS source code is available on the GRASS Web site referenced earlier. The code base is written in ANSI C programming language and is portable across common operating systems and architectures. To make the development of GIS tools more efficient, GRASS provides a large GIS library with documented application programming interfaces (API) (C and C++). Two programming models exist, namely wrapping the core GRASS modules in user-created scripts (bash shell, Python, TclTk, etc.), or directly interfacing to the GRASS API with code written in C or C++. Most repetitive or iterative tasks can be accomplished with the "script programming" approach, while more advanced users can extend existing code via the GRASS API. Several of the modules now included in the GRASS code were originally developed by users to solve highly specific research-based spatial analysis problems within research projects which were then generalized for common GIS usage.

The modular concept of GRASS is also important for facilitating further development. Most modules are also usable from the command line, which allows their integration into UNIX shell, PERL, PHP or Python scripts. The C API is exposed to other programming languages through a SWIG interface (currently PERL and Python). While GRASS provides only limited support for parallel computing (only the partial differential equations library), it is possible to use it in a simple scripted approach on clusters (Neteler et al. 2005). Significant performance gains can be accomplished through this approach, especially in image processing, where processing of image tiles can be parallelized to take advantage of multiple processor cores.

9.7.3 Room for Development: Where GRASS Needs Work

Despite the rich history and diversity of modules which make up GRASS, users have identified several weak points. Seasoned GRASS users are familiar with difficulties associated with producing publication-ready maps with GRASS, tied to the original focus on analysis by the GRASS development team. Modules such as *d.out.file* and *ps.map* can be used to make simple maps, but external vector or image editing applications are commonly employed to produce a final product (Inkscape, Skencil, etc.). The recent addition of direct PostScript output from the low-level display commands has improved this situation, however a more unified solution will probably not be integrated until the release of GRASS version 7. In the meantime, several people have discovered that publication-quality maps can be created by coupling GRASS to external, specialized applications such as Generic Mapping Tools (GMT) and MapServer. An ongoing project (until 2009) at FBK-IRST/Municipality of Trento (Italy) is aiming to develop new vector editing functionality and a graphical front-end to hard copy map production.

Automatic visualization of thematic vector data (i.e., the display of choropleth maps, with automated colour palette selection) is cited as a key feature of a GIS, and is notably missing from GRASS. Since vector analysis was added to GRASS from version 4.0, there has been a shorter amount of time for developers to implement thematic vector display. It is possible to define manually a colour palette based on individual attributes, and display each category one at a time to "build up" a final image. However, this approach is far from optimal, especially for new users. External interfaces to GRASS data such as QGIS or uDig have thematic vector display capabilities. As these applications continue to mature it is possible that they will fill the role of thematic map creation.

9.8 A Complete Geospatial Toolkit

GRASS contains nearly all of the functionality required for most types of GIS work. However, there are several cases where external applications are better suited for specialized tasks. Detailed statistical analysis, hard copy map production (as discussed earlier), or complex database queries are issues that numerous users have identified. In this context, the existing FOSS4G software stack can fill the gaps (Jolma et al. 2006). Instead of re-inventing the wheel, the GRASS development team has worked hard to leverage external, specialized applications where appropriate. Figure 9.6 illustrates some of the commonly used external tools within the context of what they are used for.

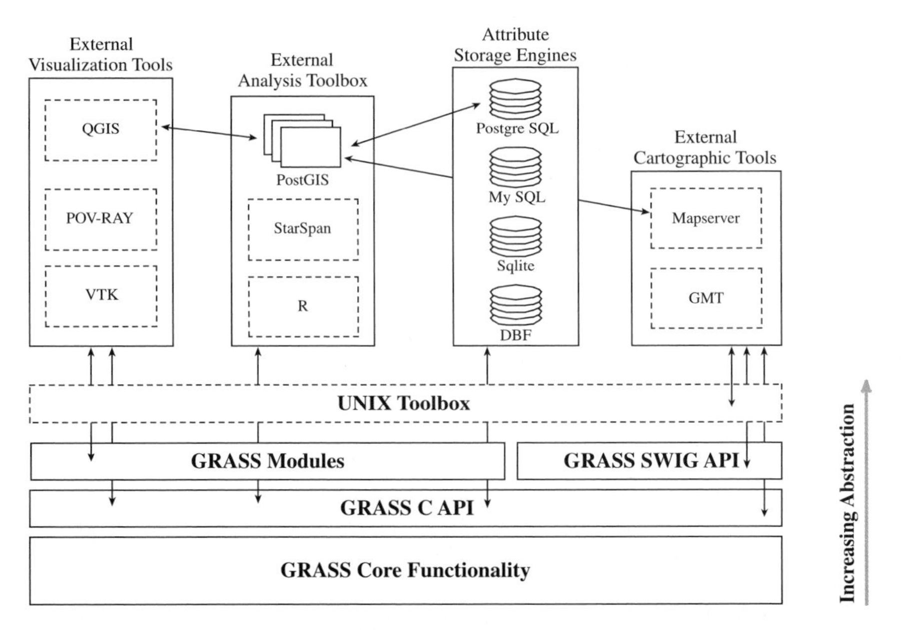

Fig. 9.6 An example of the extended GRASS geospatial toolkit: a set of visualization tools; external analysis toolbox; external attribute data management; online and press-ready map production

These applications, when coupled with the core GRASS modules through both loose (i.e., using intermediate files of commonly structured text or binary data to transfer information between applications) and tight (i.e., using program APIs and inter-process messaging to pass data between applications) coupling, can fulfill even the most demanding spatial analysis needs.

9.8.1 External Attribute Storage

The default storage engine for tabular, or attribute, data is the well-known xBASE (DBF) table format. The GRASS database abstraction routines provide a simple SQL interface to the xBASE files. However, advanced SQL constructs and data types are not supported. These are, however, available through the use of industry-strength RDBMS such as SQLite, PostgreSQL, or MySQL as the repository for attribute data. GRASS provides "tight integration" to these applications, allowing the user to switch freely between RDBMS backends with the *db.connect* module.

9.8.2 Graphical User Interfaces

Quantum GIS (QGIS, http://qgis.org) has emerged as a favourite FOSS4G 2D visualization environment for GRASS users, providing a modern user interface and map element symbology editor. The current version of QGIS contains a graphical interface to most GRASS tools, a graphical data catalogue, and a native vector digitizer. Javagrass (http://www.jgrass.org), a client-server implementation, is an alternative user interface which includes 3D visualization. It comes with special focus on hydrological and geomophological analysis. The fusion with the uDig software project is currently ongoing, adding 3D visualization and further GIS analytical capabilities to uDig.

9.8.3 Visualization

The built-in visualization module NVIZ is very efficient and powerful. It supports raster and vector data including 3D and raster volumes. Flight-simulator mode, high-resolution output, and custom animations complete the suite.

The Persistance of Vision Raytracer (POV-Ray, http://www.povray.org/) has been used for over a decade by professional graphic designers to render complex 3D scenes, complete with realistic lighting and surface interaction. The modules *r.out.pov* and *v.out.pov* convert GRASS raster and vector data types into a format suitable for inclusion in a POV-Ray scene file (Fig. 9.7(a)).

The Visualization Toolkit (VTK) has recently emerged as an exciting new way to interact with GRASS raster, vector, and volume datasets. Paraview, an integrated

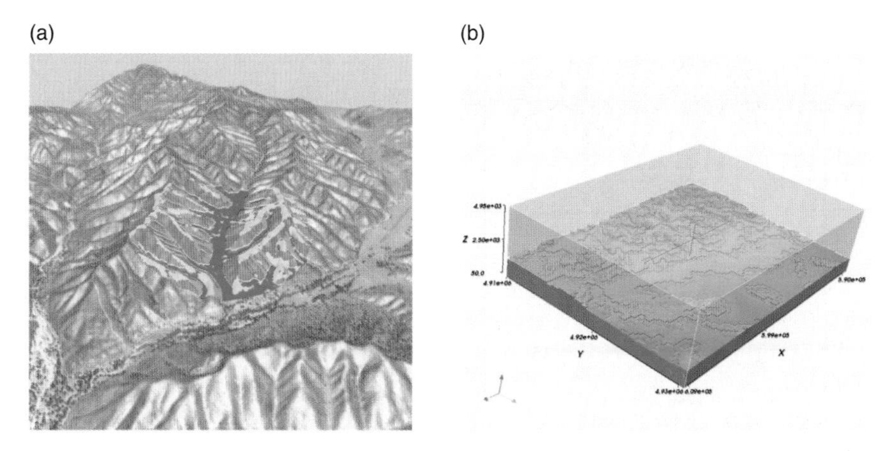

Fig. 9.7 Examples of external applications, POV-Ray (**a**) and Paraview (**b**), commonly used to visualize complex 2, 3, and 4 dimensional GRASS datasets

visualization application based on VTK, is a modern analogue to NVIZ and in its latest version is implemented with the cross-platform Qt GUI toolkit. The modules *v.out.vtk*, *r.out.vtk*, and *r3.out.vtk* are used to export GRASS raster, vector, and volume datasets (Fig. 9.7(b)).

9.8.4 Statistical Analysis

While several modules within the core GRASS system can perform simple statistical summaries or compute correlation coefficients, R (R Development Core Team 2006), the OS implementation of the "S" statistical language, is the FOSS platform of choice for detailed statistical analysis. R has a long history of compatibility with GRASS, and connecting the two is as simple as installing the necessary packages in the R environment. The interface between GRASS and R has been under rapid development over the last two years, and the current implementation has simplified the process significantly (Bivand 2007). R software is hosted at SourceForge (http://r-spatial.sourceforge.net), and includes *spgrass6* (the main R-GRASS interface), *sp* (the basic spatial object classes), *spGDAL* (a wrapper for functions in the *rgdal* package which interfaces to the GDAL library – see Chap. 5), and *spmaptools* (an interface to the *shapelib* library – see Chap. 5).

Regional or "zonal" statistics can be calculated within GRASS using the *v.rast.stats* script, however Starspan (http://starspan.casil.ucdavis.edu/), a more mature implementation of these functions written in C++, can be used instead. Starspan uses the GDAL library to read natively GRASS raster and vector data for the calculation of zonal statistics. Results are saved in comma separated values (CSV) format, which can easily be read into R for statistical analysis or back into GRASS.

Several examples of descriptive, geo-statistical, and other standard functions in R can be found in the GRASS Wiki site (http://grass.osgeo.org/wiki, see also Figs. 9.8

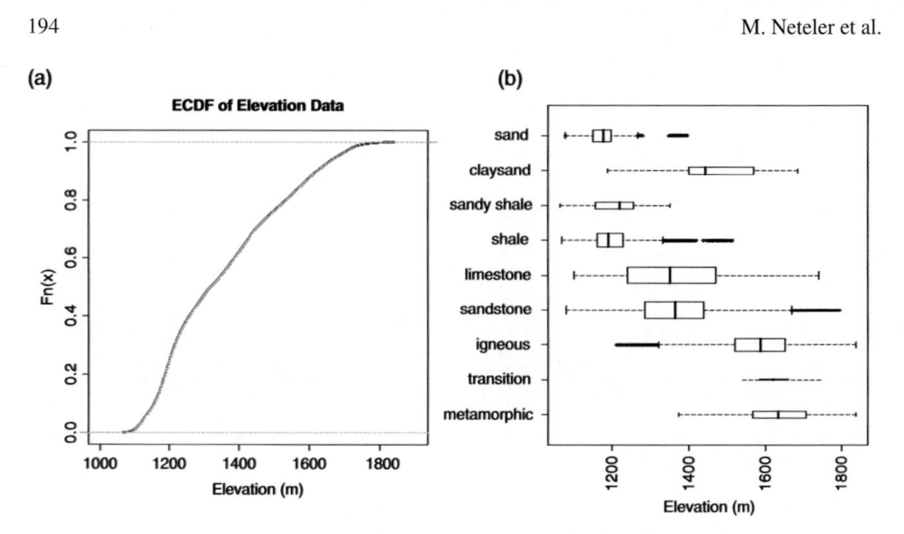

Fig. 9.8 Simple exploratory statistics performed in R on elevation (**a**) and geology (**b**) raster data

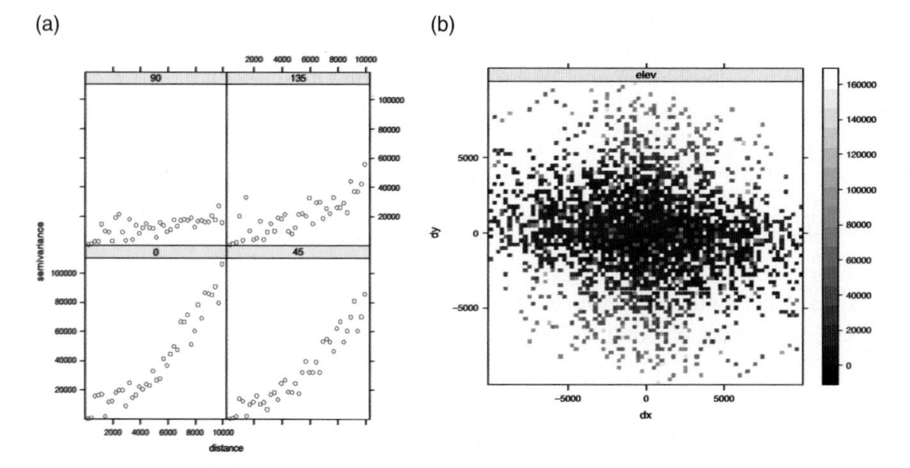

Fig. 9.9 An exploration of anisotropy with directional variograms (**a**) and a variogram map (**b**)

and 9.9). The examples used here are based on the "Spearfish" sample data location (South Dakota, USA, 103.86W, 44.49N), the classical GRASS sample data set.

9.8.5 Cartographic Tools

As noted earlier, simple maps can be printed with the help of QGIS as an interface, while several other options are available for more complex tasks. MapServer (Lime 1996 – see Chap. 4 of this book) is a common gateway interface (CGI) application targeted at the production of online maps. It functions as the primary cartographic engine behind many popular online mapping applications. MapServer can directly read GRASS raster and vector data through the GDAL library, making it possible to serve maps online immediately after they are produced without the need for intermediate conversion. The expressive map description language used by to define symbology, coupled with features such as automated label placement and collision detection make MapServer a powerful map making tool.

The GMT suite (http://gmt.soest.hawaii.edu/) consists of 60 specialized map making commands which output to publication-quality PostScript format. With GMT in hand, it is possible to fill the gap in hard copy map production which has troubled GRASS users for years. Several attempts have been made at integrating GRASS and GMT (Beaudette 2007). However, most are based on the "loosely coupled" approach of using intermediate files. Recent developments in the GDAL library, maturation of the Python/SWIG API, and planned Python integration in GMT 5.0 suggest that a more generalized and coherent fusion of GRASS and GMT will be possible in the near future.

9.9 The GRASS is Growing: Reaching the World

The source code accessibility and the modular character of GRASS render the software an ideal platform for further development. The core system with its analytical capabilities can be accessed externally using several programmable approaches in order to build graphical user interfaces upon it, or to use GRASS as the backbone for external applications.

9.9.1 Making GRASS Accessible with a GUI

In place of a monolithic graphic interface to the GRASS modules, several small GUI systems have emerged over the years. Initially based on the TclTk GUI framework *tcltkgrass*, *d.m* and its successor *gis.m* have provided a simple interface to common GIS analysis and visualization. These modules use a simple palette construct for defining a list of commands like *d.rast* or *d.vect*, with common GUI constructs such as buttons, check boxes, and forms for defining style elements. The GUI for NVIZ (the 3D interface to GRASS raster, vector, and volumetric data) and *v.digit* (the GRASS vector digitizer tool) were also built with the TclTk framework. TclTk has long been used as a robust, multi-platform GUI toolbox. However, it is starting to show its age.

The new GUI, *wxGRASS*, provides a modern-looking, compact layout of the core GRASS functionality along with operating system-native graphical elements (also known as widgets). Python/wxPython provides numerous GUI primitives (such as color pickers, combo box menus, etc.) which simplify the underlying code, and enable users to adjust display properties quickly. In addition, the conversion from

TclTK to the popular Python language opens development to a wide audience of Python programmers. Object-oriented design, simple extensions of functionality though loadable modules, and advanced array handling are just some of the features Python has to offer. The *wxGRASS* interface contains several new and updated features. A built-in attribute management system with query support simplifies table manipulation operations. Improved type rendering and printing support also facilitate rapid map production.

9.9.2 SWIG/Python Interface

The Simplified Wrapper and Interface Generator (SWIG-http://www.swig.org/) is a framework for connecting internal functions (i.e. GRASS C and C++ API) to an assortment of popular scripting languages such as TclTk, MATLAB, Perl, Python, PHP, JAVA etc. The SWIG compiler reduces the tedium of writing language-specific extension
modules by automating the entire process. A functional prototype GRASS-SWIG interface was recently implemented. The GRASS-SWIG interface opens the complex GRASS core-API functionality to developers who may be more proficient in, or prefer the flexibility of, different scripting languages. In addition, as the number of SWIG-enabled projects increases it will become possible to create a custom meta-API (in Python, for example) which bridges multiple distinct applications. More information on the GRASS-SWIG interface can be found on the GRASS Wiki.

9.9.3 GRASS: A GIS Backbone for Web Applications

Web-based spatial services are developed for many applications. In particular, opportunities for integrating data from various Spatial Data Infrastructure (SDI) portals into Internet-based services have grown. Interoperability and standardization are accomplished by following standards such as those set by the OGC for spatial data and related information technologies. While internally data storage and processing may differ from proposed OGC standards, all recent FOSS4G systems provide data exchange interfaces consistent with industrial standards. FOSS4G applications can either directly read GRASS raster and vector data through GDAL/OGR interoperability, or, GRASS vector data can be stored in a PostGIS-compliant spatial database, which is then linked to the application.

The Python Web Processing Service (PyWPS, http://pywps.wald.intevation.org) is a relatively new project started in spring 2006, which aims at implementing the OGC Web Processing Service (WPS) standard. A WPS can be configured to offer any sort of GIS functionality to clients across a network, including access to pre-programmed calculations and/or computational models that operate on spatially

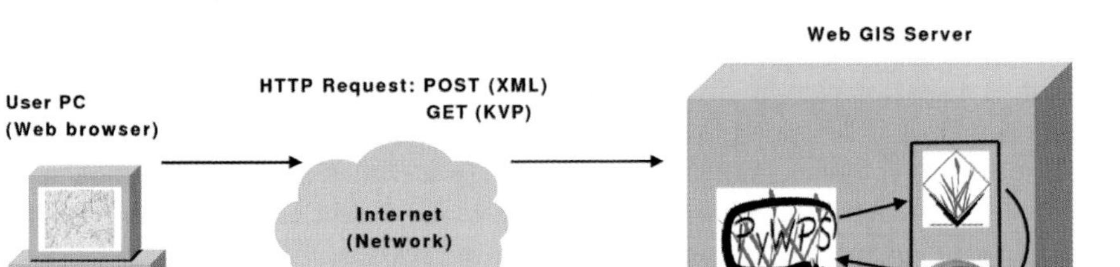

Fig. 9.10 Scheme of PyWPS (Python Web Processing Service)

referenced data. A WPS may offer calculations as simple as subtracting one set of spatially referenced numbers from another (e.g., determining the difference in influenza cases between two different seasons), or as complicated as a global climate change model. The data required by the WPS can be delivered across a network, or be available at the server.

PyWPS acts as a as translator between requests from a client and the working tool (GRASS), installed on the server. In this way, the capabilities of GRASS are extended from a simple desktop application to a complete network-oriented system. Figure 9.10 illustrates the conceptual scheme of this approach.

The WPS standard is similar to other better known OGC standards, for example the Web Mapping Service (WMS) or Web Feature Service (WCS). It accepts three types of requests, namely *GetCapabilities*, *DescribeProcess* and *Execute*. Each server offers several processes (tasks), which it is able to provide. Each process has defined inputs and outputs. Data inputs and outputs are *ComplexValue* (raster or vector maps or reference to them). Other inputs can be of type *literal value* or *bounding box value*. With GRASS, all possible tasks can be performed, which do not require direct interaction with the map display. With the combination of WPS-clients (which can be a Web application in a Web browser, or plugin to a desktop GIS), a user does not have to install large GIS packages on his/her desktop as the application is running on a remote server.

9.10 The Future: GRASS in the FOSS4G Arena

Ongoing development will lead to GRASS Version 7 which will address several important issues. The GRASS raster libraries need an overhaul in order to take advantage of modern storage mechanisms such as tiling and caching. In preparation for new (computational) challenges, parallelization of numerical operations will be extended in order to support better calculations on computer grids and distributed systems. Virtual linking of raster and vector data sources is planned, which will reduce the number of import operations for most projects. Improvements in the image processing library are directly linked to changes in the raster library. However, linking development to the Open Source Security Information Management (OSSIM) API might be a sensible way to avoid parallel development. Along with these changes, better support will be added for time series in GRASS, based on SQL constructs.

The vector processing modules are in need of transaction support. In particular, a technology is required which would preserve geometry in the event of an incomplete or otherwise unexpected termination of an editing session. Furthermore, better use of spatial indexing should reduce the computational overhead required to build or maintain topology during and after geometric operations. The net result of most of the planned changes will result in better response times, and will enable GRASS to serve better as an analytical backbone for Web GIS and Web Processing Services.

The 2007 and 2008 Google Summer of Code (SoC) program has sponsored several GRASS-related projects including improved line generalization algorithms, least-cost path calculation based on global minima, a new module to compute least-cost paths in vector space (as opposed to the current raster-based approach) and others. These projects address several long standing issues which have been raised on the GRASS mailing list and feature-request system. The SoC program will be an important developer recruitment facilitator, and future projects are already being planned.

From the end user's perspective, GRASS is one of the few complete analytical FOSS4G projects available. It has evolved from its humble origins as a land management tool for military installations into one of the most comprehensive, general purpose GIS available. Support for environmental applications has been an integral part of its twenty plus years of development. Environmental and social data from various sources can be easily integrated into a GRASS project and this makes the software an attractive option to satisfy a wide range of application needs.

References

Beaudette DE (2007) Producing press-ready maps with GRASS and GMT. J Open Source Geospatial Foundation 1:29–35

Bivand R (2007) Using the R-GRASS interface. J Open Source Geospatial Foundation 1:36–38

Blazek R (2005) Introducing the linear reference system in GRASS. Int J Geoinformatics 1(3):95–100

Bouktif S, Antoniol G, Merlo E, Neteler M (2006) A novel approach to optimize clone refactoring activity. In GECCO '06: Proceedings of the 8th annual conference on Genetic and evolutionary computation, ACM Press, New York, USA, pp 1885–1892

Brovelli M, Cannata M, Longoni U (2004) LIDAR data filtering and DTM interpolation within GRASS. Trans GIS 8(2):155–174

Hofierka J, Mitasova H, Neteler M (2008) Terrain parameterization in GRASS. In: Hengl T, Reuter H (eds) Geomorphometry: concepts, software, applications, Developments in Soil Science, Vol. 33, Elsevier, Amsterdam pp. 387–410.

Issaks EH, Srivastava RM (1989) An introduction to applied geostatistics. Oxford University Press, England

Jenson JR (2000) Remote sensing of the environment: and earth science perspective. Prentice Hall, New Jersey

Jolma A, Ames D, Horning N, Neteler M, Racicot A, Sutton T (2006) Free and open source geospatial tools for environmental modeling and management. In: Voinov A (ed) Proc. iEMSs 2006, Session W13, July 9–13, 2006, Burlington, Vermont, USA

Lime S (1996) UMN MapServer, University of Minnesota, USA, [Computer Program]. Available: http://mapserver.gis.umn.edu

Lo CP, Yeung AKW (2006) Concepts and techniques of geographic information systems. Prentice Hall, New Jersey

Mitasova H, Hofierka J (1993) Interpolation by regularized spline with tension: II. Application to terrain modeling and surface geometry analysis. Math Geol 25(6):657–669

Mitasova H, Mitas L (1993) Interpolation by regularized spline with tension: I. Theory and implementation. Math Geol 25(6):641–655

Neteler M, Grasso D, Michelazzi I, Miori L, Merler S, Furlanello C (2005) An integrated toolbox for image registration, fusion and classification. Int J Geoinformatics 1:51–61

Neteler M, Mitasova H (2008) Open source GIS: A GRASS GIS approach. 3 edn. Springer, New York

R Development Core Team (2006) R: A language and environment for statistical computing. R foundation for statistical computing, Vienna, Austria. ISBN 3-900051-07-0

Rizzoli A, Neteler M, Rosà R, Versini W, Cristofolini A, Bregoli M, Buckley A, Gould E (2007) Early detection of TBEv spatial distribution and activity in the Province of Trento assessed using serological and remotely-sensed climatic data. Geospatial Health 1(2):169–176

Chapter 10
GeoVISTA *Studio*: Reusability by Design

Mark Gahegan, Frank Hardisty, Urška Demšar and Masa Takatsuka

Abstract This chapter describes the need for an open source problem-solving environment for the geospatial community, and the efforts to meet this need with GeoVISTA *Studio*, an Open Source Java environment that provides tools for geovisualisation, knowledge discovery and analysis. The four-part architecture of *Studio* is presented, along with two case studies from *Studio* users. The conclusions summarize how and to what extent *Studio* meets its design criteria.

10.1 *Studio* and the Need for Open Systems

"Systems, scientific and philosophic, come and go. Each method of limited understanding is at length exhausted. In its prime each system is a triumphant success: in its decay it is an obstructive nuisance."

(Alfred North Whitehead—Adventures of Ideas: 1933)

GeoVISTA *Studio* contains a suite of Free and Open Source Software for Geospatial (FOSS4G) components, coupled with an environment to connect them together to build and deploy a variety of applications. This environment allows users to build applications for geovisualization, geocomputation, and knowledge discovery using a visual programming editor. Unlike conventional programming, software components are connected together using a drag-and-drop visual editor to construct new

Mark Gahegan
Department of Geography, University of Auckland, Private Bag, Auckland, New Zealand,
e-mail: mark@geog.psu.edu

Frank Hardisty
GeoVISTA Center, Department of Geography, Penn State University, Pennsylvania, USA,
e-mail: hardisty@psu.edu

Urška Demšar
National Centre for Geocomputation, National University of Ireland, Ireland,
e-mail: urska.demsar@nuim.ie

Masa Takatsuka
ViSLAB, Information Technologies, University of Sydney, Australia,
e-mail: masa@vislab.net

G.B. Hall, M.G. Leahy (eds.), *Open Source Approaches in Spatial Data Handling.*
Advances in Geographic Information Science 2,

applications (Gahegan et al. 2002). *Studio* uses a component-based software model, where functionality is implemented using JavaBeans, and where new components can easily be imported and assimilated. *Studio* is a relatively long standing FOSS4G project, first released in 2000, that has been used in many applications, from land-cover analysis to epidemiology. It is produced by the GeoVISTA Centre at Penn State University.

Studio is really a problem solving environment (Chin et al. 2001). It consists of a toolkit from which solutions to specific problems can be quickly created with a minimum of programming effort. Matlab (http://www.mathworks.com/) and IBM's Open Data Explorer (http://www.opendx.org/) for mathematics and information visualization, respectively, are examples that can be considered in the same way. Importantly, *Studio* is a toolkit to which it is very easy to add new tools. The cost of integrating new functionality is very low (sometimes zero) in terms of customizing or re-engineering. As is shown below, *Studio* can readily incorporate content without the need to adapt to particular data models or interface specifications, or even the need to agree on a specific computational platform. The only hard constraint is that any methods to be added must be written in Java (any recent version of Java should suffice).

This chapter begins with a review of some of the software problems and challenges facing developers of FOSS4G projects, from both a practical and philosophical perspective. Then, in Sect. 10.2, eight specific challenges are set out that *Studio* is designed to address. Next, the architecture of *Studio* is described in Sect. 10.3, under the headings of (i) the software library, (ii) the *Studio* Engine, (iii) data models, and (iv) coordination models. Section 10.4 then gives two use-cases of *Studio*, with commentary from the perspective of a developer and a user of the project. Section 10.5 concludes by revisiting the initial challenges and describing some future goals that the developers are currently working towards.

10.1.1 Software Challenges in GIScience

The GIScience community (along with many other science communities) faces some difficult challenges regarding the tools and analysis methods that empower geographical analysis and modelling (Bishr 1998). Three of the major challenges are as follows:

1. Although it is still difficult to share data in a meaningful way between researchers and systems, it is even more difficult to share functionality. Many potentially useful tools for spatial analysis and modelling, visualization, spatial reasoning, geocomputation, and so forth are not portable in the sense that they cannot be readily shared with others. As a result, research outcomes have reduced impact both within and outside of academia, and communities often end up implementing the same idea many times.
2. Perhaps even more importantly, the inability to share methods effectively (Goodchild 2000) means that the results of analysis cannot be independently

verified or refuted by other scientists. This is a real barrier to scientific progress and credibility.
3. A richer toolbox of methods is only useful if ways can be developed to combine, reconfigure, test, and use methods that are flexible, interactive, and easily deployable (Griss 2000).

The Earth is an open system, but software systems are closed. Bearing in mind the opening quote from Whitehead, software developers must consider whether their systems act as a barrier to innovation and conceptual shift, which (should) follow from their science. A cursory look at geographic information systems (GIS) might lead to the conclusion that there is little about the world that is left to discover, because these systems (i) do not support discovery well, and (ii) do not adapt well to new discoveries and conceptual shifts. However, GIS are easy to customize in simple ways by adding or removing functionality, changing the user interface, and, with some additional effort, introducing new analysis methods. However, in deeper ways current systems are very difficult to extend, and thus lock both developers and users into certain ways of thinking. Imagine trying to add to a GIS a new data model that provides full support for time (Peuquet 1994), 3D topology (Ellul and Haklay 2006), spatial analysis based around Voronoi polygons rather than gridcells (Okabe et al. 2000), or social critiques (Gahegan and Pike 2006). These additions would be difficult to make, and likely would be very inefficient without considerable knowledge of, and access to, the inner workings (source code) of commercial systems. GIS vendors have, quite reasonably, decided what constitutes a useful set of possibilities for extending or customizing their products, and while this may be useful for many applications, it can also act as a barrier to fundamental shifts and progress in science.

If the Earth is an open system, new discoveries are possible, and there is the need, from time to time, to rethink conceptual models, then why do software systems remain closed? As long as our understanding of the Earth is incomplete, geographers will remain engaged in the process of discovery and in the creation of new conceptual models. These conceptual models will require new computational systems, or much more flexible versions of the systems that are currently in use. It is even possible to conclude, more strongly, that not to plan for evolution in our systems is a serious impediment to the progress of science.

Moving forward requires the following steps: (i) include in GIS the tools that empower exploration and discovery, (ii) design GIS so that they can accommodate new methods, models, and theories (i.e. systems that can evolve), and (iii) in an evolving computational system, make explicit the methods, models, and theories that are used as these are subject to change.

In conclusion, many of the systems in use can only be extended via proprietary interfaces which provide limited access to certain software functionality. As a consequence, independent development and deployment of new tools is possible only within the fixed bounds of a particular environment defined by its application programming interface (API). This approach acts as a barrier to research, locking users into specific ways of working and, more importantly, thinking.

10.1.2 Software, Languages, Paradigms and Their Limitations

This section reviews some of the history of computing languages and engineering paradigms. This will show that part of the problem that developers face in making systems more open and more extensible is an indirect consequence of the legacy engineering used in many of the current systems.

The computer industry has moved through several different models for systems engineering, including the procedural languages of the 1950s–80s (e.g. Fortran, Cobol, Pascal, C), the object-oriented languages of the 1980s–90s (C++, ADA), and now component-oriented approaches such as Java and C# (Szyperski 1997). Programming languages are a means of abstracting the functionality of a computing platform to a level that optimizes programmer productivity, program re-use and maintenance, and flexibility. The deeper understanding of the software life cycle developed over the last 30 years has led to progressive refinements and sometimes radical shifts in the theory of programming languages and their subsequent use in software development.

10.1.3 Component Engineering Technologies

The latest component-based languages were developed to extend the capabilities of object-oriented programming into a networked, collaborative environment where individual pieces of functionality can be deployed, assembled, and used together without the requirement that they are compiled together or even designed to work together in the first place (Sametinger 1997). Such an approach is known as late-binding, in which components are not necessarily assembled into an application until they are about to be used (run-time). This kind of flexibility is proving very useful in the development of Web Services, which in most cases must work as distinct components in a workflow without any knowledge of which components come before or after them in a processing chain. A component approach differs subtly from an object-oriented method in that it also promotes or advertises its interfaces, while hiding its inner workings. It can usually be deployed as a separate executable module which can be utilized simply by understanding how to connect to it. In terms of data structures and data modelling, the component model does not extend the basic expressiveness of the object-oriented model. However, it does change the flexibility of use, and enhances the collaboration possibilities between developers since it becomes much easier to import and exchange components (Wills 2001).

The newer component-oriented languages, as exemplified by Java via JavaBeans™, provide containers for components or classes as a convenient way to share collections of more primitive functions or objects through a process known as encapsulation. These languages offer four improvements and advantages for the developer and user of scientific systems. First, as previously mentioned, components can be distributed and used without an understanding of their internal structures and algorithms. Only knowledge of the interface is required, which can ease the burden on both the developer and the user of the software. Second, in the case of JavaBeans, introspection (a language feature) can provide details of supported interfaces even if the source code for a Bean is not available, or if the owners of the intellectual property (IP) do not wish to divulge the details. This offers the advantage of enabling sharing without the necessity of giving away secrets or knowledge, and also helps protect IP where it is appropriate to do so. Third, component behaviour can be modified and extended without requiring access to the original source code. New methods can simply be replaced or added so that, for example, existing spatial analysis tools can be modified or extended by the wider user community. Finally, sets of components can be packaged together into applications or applets that can be easily distributed or even accessed via Web browsers or as Web services, avoiding the need for end-users to own a copy of the original 'system'.

As GIScience researchers begin to embrace the newer paradigm of component-oriented engineering, it becomes clear that the advantages on offer could have significant, positive impacts on the way in which researchers design, develop, and share software tools. Not least among these is the ability for an entire research community to collaborate in the development of useful functionality using a shared software repository, source-code management tools, and well-defined interfaces.

However, component programming within a community brings into sharp focus the problems related to the conceptual models, data structures, and ontological commitments held by different developers, and the way that these are reflected in the components they develop. Components drawn from a single source (such as a commercial GIS or from a single researcher or lab) may well work together in a consistent manner, since those involved in constructing the components share a consistent vision of how they might be used together. However, when divorced from this context, it is often unclear how such components should be used, or how they should be connected together. The problem of connecting components together has dimensions that extend beyond the syntactic and schematic levels, to the semantic level. Thus, issues of conceptual models (of geographical space), data models (for passing data around), and coordination or control mechanisms (governing the flow and interaction between components) need to be considered and resolved. While not addressed here, these problems are severe enough to offset most of the potential advantages component-oriented software has to offer, and without careful consideration and additional research (e.g. O'Brien and Gahegan 2004), the substantial benefits of component engineering could completely elude the GIScience community.

10.2 The Motivations and History of GeoVISTA *Studio*

Development of *Studio* began in 1999 as an attempt to address a number of problems and shortcomings that Takatsuka and Gahegan (2002) had noticed with supporting the needs of a research laboratory (such as the GeoVISTA Centre: (http://www.geovista.psu.edu), to which Takatsuka and Gahegan had recently relocated. Some of these problems are very practical, while others are quite theoretical.

The final section of this chapter summarises how much success was realized in addressing these problems. Many of the problems derive from the problem description in Sect. 10.1:

1. Current systems to support science are not extensible enough - they limit thinking.
2. It is difficult for researchers to share functionality, and it is expensive to re-engineer continually the work of others so that it fits into a given system.
3. As scientists, there is a need to get better at remembering the details of workflows that are created and used to for analysis and modelling.
4. Following from point 3, most science applications and workflows are difficult to share with other researchers in a complete (encapsulated) way, so that results can be independently verified (or contested).
5. It is very difficult for non-programmers to make use of complex functionality, unless the complexities can be hidden or abstracted away.
6. Current GIS are not well suited to many important aspects of geographical research, such as exploratory visualization, knowledge discovery, and generally dealing with highly-multivariate data.
7. When code developers (graduate students, post-doctoral researchers, and independent software developers) leave, their code usually 'dies', as others do not know enough about it to maintain or run it.
8. New systems, once created, need to be easily deployed (i.e., turned into stand-alone, executable applications that can be run with little or no additional configuration or systems programming).

Studio has grown in the past seven years, sometimes erratically, into a comprehensive geovisualization environment, with good support for knowledge discovery. However, it is quite limited in terms of support for more traditional GIS and spatial analysis functionality. Its content was developed to suit specific questions pursued by the researchers that comprised its development team, or in response to funded research projects. Content was also added from external sources on occasion, including a light-weight parallel coordinate plot, some cartographic tools, and a statistical library. It was never the intention to recreate an existing GIS, but rather to concentrate the component development effort on areas not covered well in existing GIS. However, components derived from the *GeoTools* open source project (http://geotools.codehaus.org/) have recently started to be added to *Studio*, since *GeoTools* is based on open standards and provides useful functionality to input, filter, and display geographical information (see Chap. 8 of this book). A mark of success (point 7 above) is that *Studio* has survived all of its initial developers leaving.

10.3 The *Studio* Architecture and Component Library

This section describes the basic architecture of *Studio*, and some of the component collections that can be used to construct applications via a workflow or 'design'. Some of the tool-sets that were created explicitly for *Studio* are presented, along with other components that were acquired from other (sometimes non-geographical) OS communities. *Studio* comprises four completely separate entities: (i) the software library, (ii) the *Studio* Engine, (iii) a data model, and (iv) a coordination model.

10.3.1 The Software Library

Studio's software library is comprised of components implemented as JavaBeans. The library contains a mix of functionality created by the developers, and components harvested from other OS projects (where licenses allow). For neatness, the components are organized into palettes, which can be customized and reorganized by the user, if needed. Users may also add components into palettes via a simple registration process. Table 10.1 shows a subset of the palettes currently distributed. Note that the palettes contain components for control and simple interactions (the swing package), visualization (GeoViz and Java3D), analysis (the self organizing map (SOM) and spatial data mining palettes), and interaction with the *GeoTools* open source project (and hence with open geospatial consortium (OGC) standards for geographical features and map styling).

10.3.2 The Studio *Engine*

The Engine, at the core of *Studio*, provides an execution environment for connecting components together, an editor for visually creating workflows, and the mechanisms to save, load, and execute these workflows. The *Studio* engine is built on the Java 2 Platform (i.e., above the Java Virtual Machine (JVM), which provides platform independence), and below the JavaBean API. Thus, it can utilize the latter to help create and manage workflows. Figure 10.1 shows this arrangement, where Beans accessed via the JavaBean API are connected together into an executable sequence.

Users interact with a visual editor to construct their workflows in *Studio*, by dragging components that are organized into palettes onto the design canvas. Figure 10.2 shows a design that has been annotated to explain what the various pieces do. Components are connected together by literally drawing a line between them in the visual editor. Introspection is performed automatically on the components to be connected, and their methods are exposed to the user. The user must then select which method is to be used in a given connection. In this manner, *Studio* avoids the need for the user to become encumbered by issues of programming and syntax, although the user must still know which of the methods (exposed as ports) supported by a component should be used in a connection—with the typical solution being to use an event callback method, where an event (in the programming sense) in the sending (output) component causes the receiving (input) component to request data (Takatsuka and Gahegan 2002). However, the weakness with this method is that the designer needs to know which, from a set of possible ports, are the appropriate ones to connect

Table 10.1 A selection of some of the *Studio* component palettes which are available for use in constructing a 'design'

Palette	Domains of Origin	Sample JavaBeans Components
Swing	Basic GUI components from the javax.swing package.	
GeoViz	General components for geographical visualization.	
Java3D	3D visualization components based on Java3D technology.	
SOM	Pattern clustering and classification tools based on self-organized map.	
Spatial Data Mining	Specialized data analysis and information visualization tools tailored for spatial data mining.	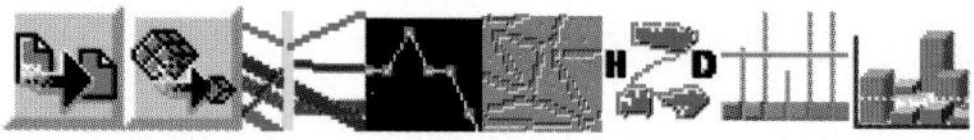
Geo-Tools	Geospatial data handling tools based on OGC standards and the GeoTools API.	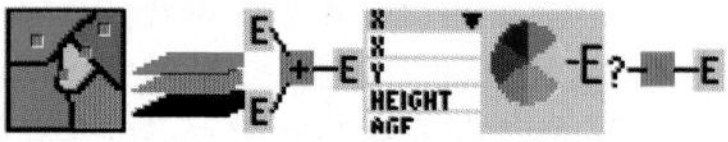

to. Naming conventions can help to make this choice clear in many circumstances, as does the creation of an automatic coordinator (see below), which reduces the combinatorial explosion of connections between components working together in a tightly orchestrated manner. However, in some cases the need for at least some programming-level knowledge cannot be avoided in order to understand fully how to create designs in *Studio*.

The issue here is one of abstraction, in that the more flexible and general a system tries to be, the less its users can rely on simple workflows and development strategies that work only in a narrow set of circumstances. In this debate, *Studio* tends to err in favour of flexibility and generality.

The 'design' that a user produces can be considered as a live experiment, since execution begins as soon as data are imported. Users can modify the design interactively, without stopping to reload data or perform any housekeeping. Thus,

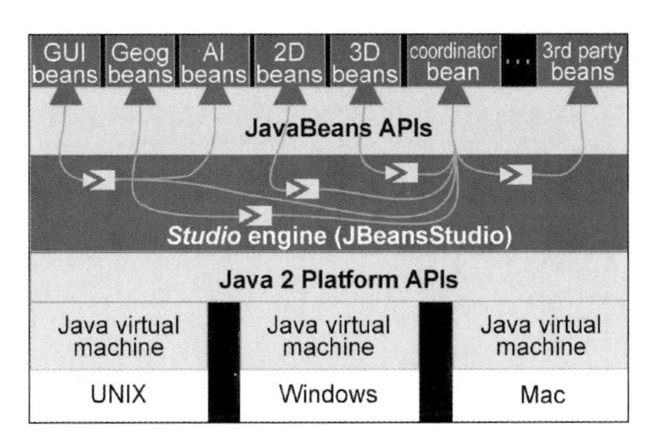

Fig. 10.1 Overview of the *Studio* architecture (darker layers represent *Studio*, lighter layers represent aspects of the Java environment)

if during some analysis of a complex dataset the user needs to incorporate additional or alternative visualization components, or to cluster the data, the user can simply 'drop in' the new components from one of the palettes in the design window, connect the component using appropriate links, and continue with the analysis.

As a design is constructed, each component that is added shows its visual interface on a separate display window. The image in Fig. 10.3 shows the display created for the exploration of a point dataset of soil samples and their properties. It contains an image renderer, a scatterplot matrix, a parallel coordinate plot, a spreadsheet, and legend controls (that use the OGC's Styled Layer Description layout standards).

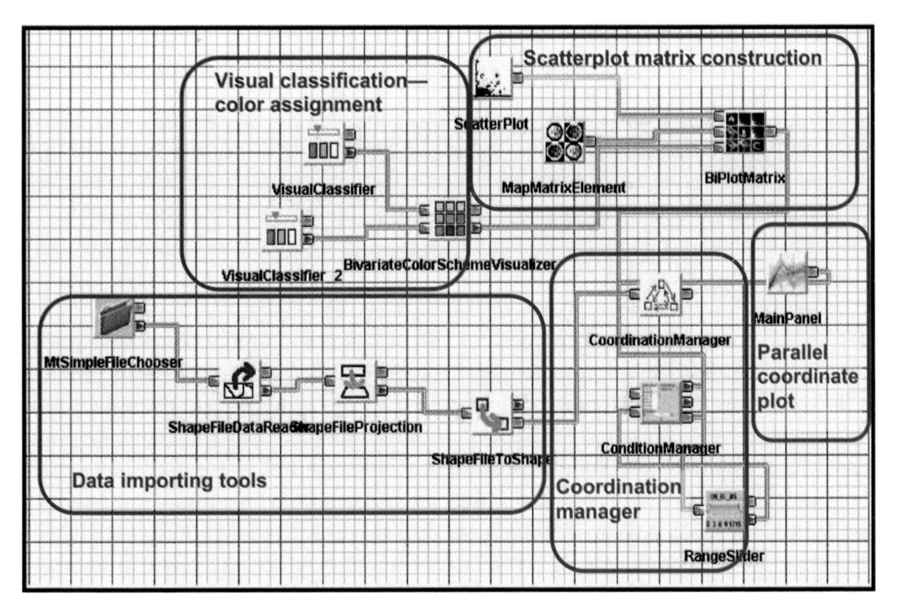

Fig. 10.2 A *Studio* design window, showing a design to create a coordinated display of geographical information between several visualization components. The figure is annotated to highlight the roles of each component

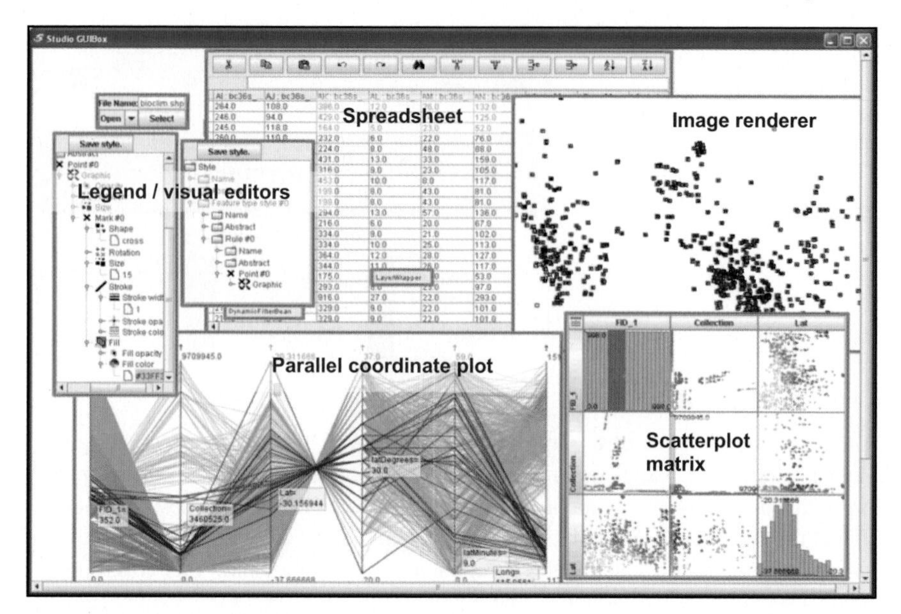

Fig. 10.3 The *Studio* visualization window, where the visual interface of each component used in a design is displayed for the user to interact with. The dataset on display shows the position and properties of soil samples

10.3.3 The Data Model

The studio data model establishes a data structure and set of interfaces to allow spatial and non-spatial data to be passed expediently between components. Having any kind of data model does enforce certain constraints in terms of representation and functionality since it defines how data will be presented to the components. It should be emphasized that the use of any data model is optional, and simply a matter of developer convenience.

Currently, *Studio* offers a choice of data models, including a lightweight model called DataSetForApps, used by most of the visualization components developed so far, and the GeoTools implementation of the OGC's (Simple) Feature Model (http://www.opengeospatial.org/standards/sfa), which provides an elaborate and comprehensive representation of spatial data and support for backend databases and query filters (complex, editable expressions). Using both of these models (or any other models) concurrently is possible since *Studio* supports the notion of *adaptors*. An adaptor is a special class of component that provides a translation service when the output of one component is not of the required type for the input to another component. Typically, adaptors are hidden in *Studio* designs, and in some cases can be automatically deployed where needed in a workflow.

The design of a suitable data model has been a source of great controversy among the developers, with most favouring the lightweight approach because it is easy to develop against a simple interface. However, complex and voluminous data require more sophisticated support. Thus some of the more recent data models designed specifically to support information visualization tasks are currently under review (e.g. *prefuse*: http://prefuse.org/), with a view to adopting one of them in addition to the above. Adaptors will then be created to allow existing components, developed against existing data models, to continue to be used.

10.3.4 The Coordination Model

Coordination is essential for many kinds of applications where user interactions need to be shared and echoed throughout a set of components that are being used together. In visualization parlance, such behaviour is often known as 'linking and brushing'. A typical example is when the user highlights a region in a scatterplot (e.g. by dragging a cursor across the display) and the selected points are also highlighted in a parallel coordinate plot, or a map. Coordination can be achieved in a variety of ways, but it essentially involves providing a mechanism where certain components can listen for and respond to events generated by other components. So if a scatterplot can generate an event such as 'a new selection has just been made', and a map can listen for that event, then it can take a similar action, or translate that action into some other event. The *Studio* design shown in Fig. 10.3 shows components for supporting coordination wired into the workflow. Similar to the data model, users are free to substitute their own coordination strategies, or to ignore coordination completely.

A significant advantage of using a coordination model is that it greatly simplifies the number and types of connections that a developer must create between components in a design. This is because the coordinator acts as a kind of agreed specification for listening for, broadcasting, and responding to different kinds of events that are typically generated by user-interaction with visualization tools. Figure 10.4 shows two designs that achieve the same level of coordination. The

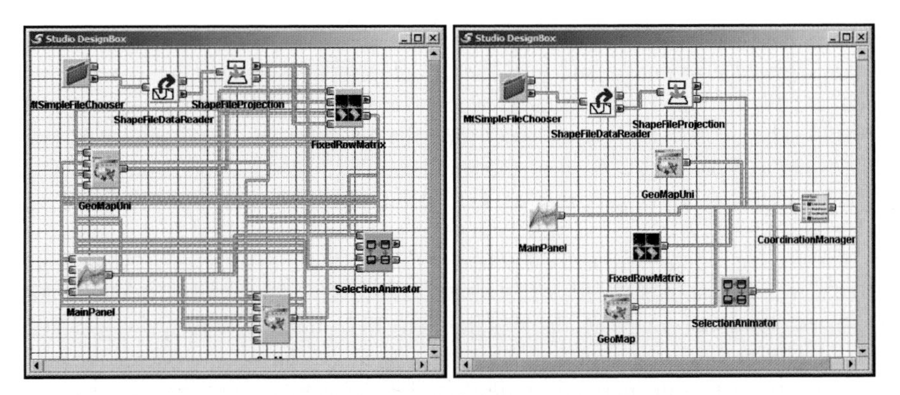

Fig. 10.4 Explicit designs (*left*) vs. coordinator simplified designs (*right*)

left design uses explicit connections between components, while the right design shows a much-simplified topology that is produced after the Coordination Manager component is introduced into the workflow. The remaining connections are simply used to pass data. Further details of the coordination strategy can be found in Hardisty (2002).

10.4 Some Examples of Studio in Use and Spin-offs

Studio has provided support for several research projects and sponsored activities. This section showcases two different experiences of using (and developing) *Studio*. The first is from Urška Demšar, whose doctoral dissertation examined the ways in which visual and computational components can be integrated to explore and analyze complex data. The second is from Frank Hardisty, whose doctoral dissertation revolved around research conducted for developing and coordinating visualization components.

10.4.1 Urška Demšar: Studio *User*

The following is a personal account of how GeoVISTA *Studio* was learnt and used by a non-developer. There are two aspects to consider when beginning to use *Studio*. The first is how to build custom exploration systems (designs) using the visual programming interface, and the second is how to then use this custom-built system for the actual task of data exploration.

Initially, the idea of visual programming in *Studio* may seem appealing and easy to learn. In reality, however, there is a rather steep learning curve involved. One of the reasons for this is that since *Studio* was developed as an academic project, at the time of writing there is no comprehensive guide for beginners available. The *Studio* website provides a Quick Start Guide, a number of excellent tutorials (covering how to use your own data, and several particular applications including ESTAT, data mining, and matrices), some examples, and a list of published Java Beans. However, to understand fully the design principles and implementation, as well as how particular visualizations work, it is essential to read the papers published by the GeoVISTA Centre, which describe the different components of *Studio* (for example, Takatsuka 2001; Dai and Hardisty 2002; Gahegan and Brodaric 2002; MacEachren et al. 2003; Edsall 2003; Guo et al. 2004; Robinson et al. 2005 among others).

The necessary information for a potential system designer is to some extent scattered across these publications, and requires determination to process, though there are many additional papers available in the GeoVISTA library (http://www.geovista.psu.edu/dl/library/CitationFull.jsp). Once the system designer has mastered the process of constructing the network architecture by connecting appropriate Beans, and has become familiar with various components, building a custom system for each new application becomes fairly straightforward.

Using a custom-built system for data exploration also promises to be a relatively easy task to master. Again, once the user has learnt how to manipulate and interact with various visualizations, it is in fact easy to explore the data and identify various patterns. However, experience shows that visually exploring data in interactive connected multiple views is a foreign or novel concept for many users. While the whole principle might seem very confusing to a beginner, usability experiments show that most users are able to grasp the concept fairly quickly (Demšar 2006, 2007b).

Nevertheless, significant interpersonal differences do exist, related not only to an individual's learning and perceptual abilities, but also to the complexity of visualizations. Humans are good at visual pattern recognition, but it is difficult for certain individuals to translate a pattern that they see in an abstract display into a meaningful observation about the data. This is a known issue in information visualization (Plaisant 2004). It is especially true for complex visualizations that the user is unfamiliar with, particularly if the pattern is not very conspicuous, as is often the case with patterns in real data. Several of the displays in *Studio* fall into this category, such as the space-filling technique for ordering data records in a regular space by their value for one attribute, while colouring them based on another attribute value.

As an example, Fig. 10.5 shows a multiform bivariate matrix from *Studio* displaying (left) a synthetic spatial dataset (Demšar 2006) based on the famous iris data (Fisher 1936) and (right) a real environmental dataset (Demšar 2007a). Patterns in the synthetic data are more striking than in the real data. Notice the linear separability of two clusters in several scatterplots on the left and relatively smooth white-grey-black vertical ordering in several of the left space-fills. In contrast, scatterplots

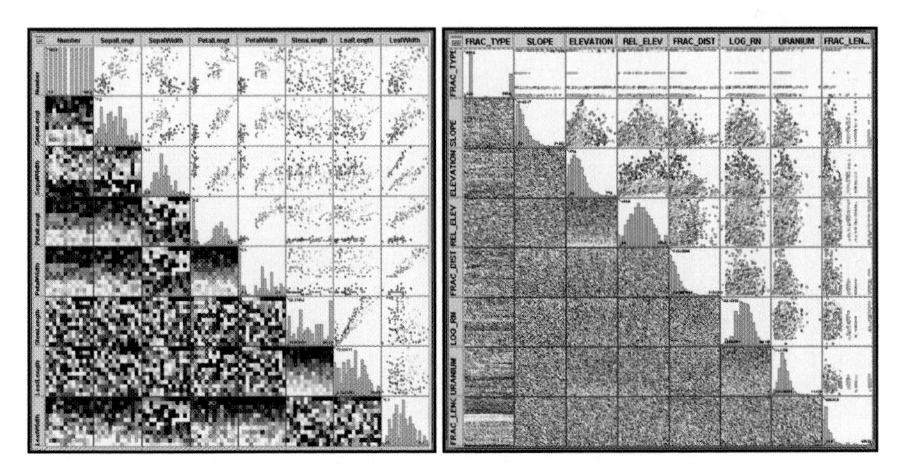

Fig. 10.5 The multiform bivariate matrices from *Studio*, showing synthetic (*left*) and real (*right*) data

of numerical variables on the right show a much more random scattering and only a few of the right space-fills have visible ordering.

An example of a useful first insight discovery is demonstrated in a study by Demšar (2007a), where a spatial cluster was identified by visual analysis of a non-spatial display without any reference to the geographic location of the data. This cluster turned out to be of particular importance for an environmental research group that this study was designed for, since the identified cluster was located in an area with atypical environmental conditions for the phenomenon under investigation (Fig. 10.6, right). The visual identification of this pattern indicated commonalities between data in this particular area and this was considered to be important enough to warrant further environmental investigation.

Visual exploration with *Studio* tools can also be useful as a post-analysis step for interpretation of results of a spatial statistical method. Demšar et al. (2007), for example present a case where the multivariate parameter estimate space of

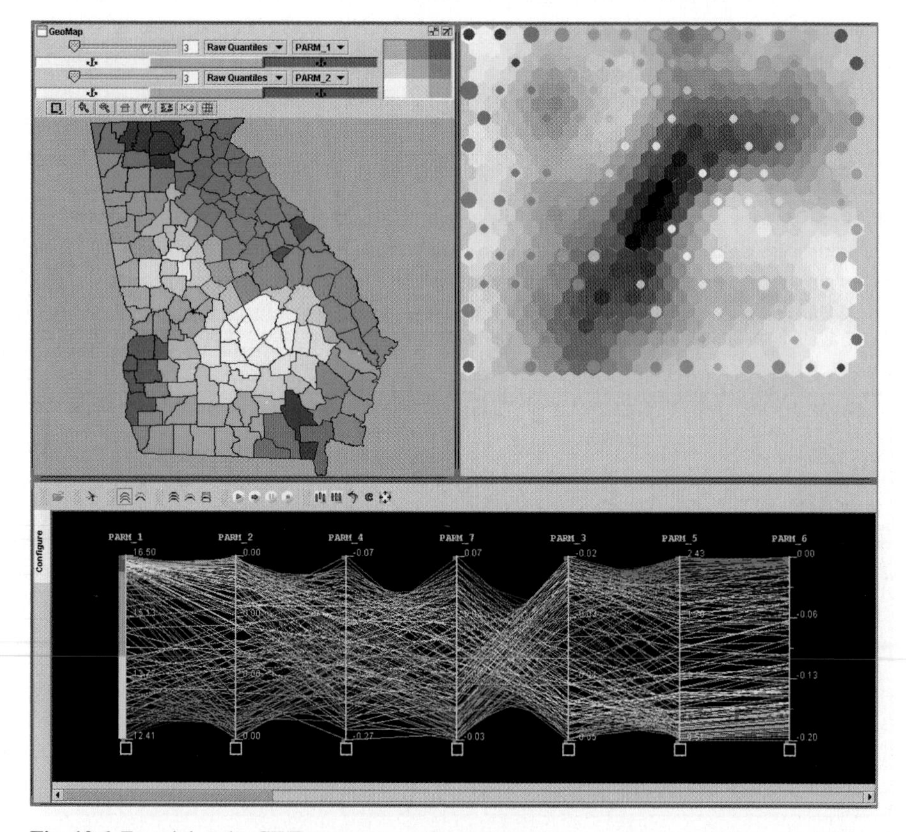

Fig. 10.6 Examining the GWR parameter estimate space in an exploratory environment built using GeoVISTA *Studio*. Components shown are a thematic map (*top left*), parallel coordinate plot (*bottom*), and a self-organizing map for clustering (*top right*)

Geographically Weighted Regression (GWR) was explored using *Studio*. Traditional regression analysis describes a modelled relationship between a dependent variable and a set of independent variables, where a parameter belonging to each variable is assumed to be stationary over the whole area. GWR extends this method by dropping the stationarity assumption, whereby the parameters are assumed to be continuous functions of location (Fig. 10.6).

GWR results are continuous localised parameter estimate surfaces, which are traditionally visualised as univariate maps and compared to each other. This works well if each parameter surface is to be investigated separately, but not so well if the analyst wishes to answer questions such as 'Do areas of stability exist where all the parameters keep relatively constant values?', or 'Are there any predominant groupings of parameters that behave in a similar way everywhere in the area of investigation?' *Studio*-based tools provide a means to regard the parameter estimates as a multivariate dataset, and to identify spatial and multivariate patterns, such as those shown in Fig. 10.6, that describe the spatial variability of the parameters and underlying spatial processes.

Once the initial learning steps have been mastered, *Studio*-based systems become powerful tools for data exploration, and can be applied for a variety of purposes. In the two examples presented here, *Studio* was used for two very different tasks, namely to gain initial insights into a dataset and to facilitate the interpretation of results of a traditional data analysis method.

10.4.2 Frank Hardisty: Studio *Developer*

This section is written from the perspective of a developer of geographic visualization tools who contributed components to *Studio*. *Studio* is a successful project in that it has advanced the state of the art for coordinated run-time geographic visualizations (see Fig. 10.7), and provided a platform for the development of a variety of fully-realized geographic visualization and analysis strategies, which still work today. The basic philosophy of *Studio* (run-time coordination of separately instantiated components) has been validated by the ability to re-use geographic visualization and analysis tools in a variety of contexts over a period of several years. Its strengths of flexibility and openness lead to some complexities for users and programmers alike, so it is not for everyone.

One of the key reasons for developing *Studio* was that research projects too often end up orphaned. There is often no way to test or observe firsthand how a given piece of software works. At the GeoVISTA Centre, and at other university labs, work by many graduate students has been effectively lost because of the difficulty involved for others to rebuild and review their efforts. While the conceptual contributions remain valid, there is value in being able to compare the actual working versions of technology projects and reuse their content. *Studio* provides

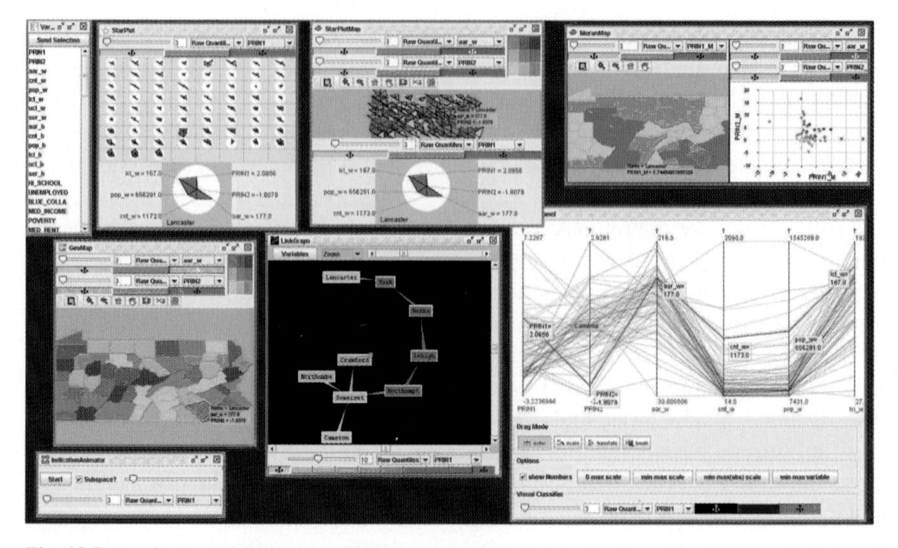

Fig. 10.7 A selection of *Studio* visualization components, many developed by Hardisty, being used together in the analysis of demographic data for the state of Pennsylvania

a focus for many research projects which, having been developed in accordance with *Studio*'s principles, stand a much better chance of being useful today. Finally, while *Studio* was initially developed some time ago, it is still unsurpassed in some ways. A prime example of this is that *Studio* allows users to assemble visually a novel workflow, which can be compiled and distributed as a new lightweight executable application. This is an impressive accomplishment for any technology.

From a programmer's perspective, contributing to a component-based problem-solving environment represents a shift in thinking from creating a stand-alone application to serve only the developer's own needs, to designing (and exposing) appropriate methods *and event mechanisms* so that components can play a variety of roles in systems not yet constructed and for problems not yet envisaged. In other words, much development takes the form of designing, building, and testing libraries of components, often within a group development setting. Most graduate students interested in GIScience have some programming experience, but perhaps not enough to prepare them for this kind of software development. Thus, all of the *Studio* core code, and most of the application code, was ultimately created by a smaller group of researchers.

As a side note, software development cycles are shortening, and there has been an explosion in both the quality and quantity of OS software available. However, the learning curve for new programmers wanting to contribute useful tools is steepening, rather than becoming easier. One of the abiding challenges for FOSS4G, which *Studio* partially addresses, is to find ways to allow those without deep programming expertise to contribute to geographic analysis research.

10.5 Current and Future Directions for *Studio*

This section describes some of the latest developments and ideas for the future of geographical problem solving environments, covering spin-off projects from the Studio code-base, creating Web-applications and services, constructing bridges to other applications and the addition of automated reasoning to help users create designs more easily.

10.5.1 New Systems for Geovisualization

One of the lasting benefits of *Studio* has been the development of a code-base that has a high degree of internal consistency. This allows the components to be used together with relative ease. As a consequence, the code-base is drawn on extensively to support the development of additional stand-alone applications, including ESTAT (http://www.geovista.psu.edu/ESTAT/), which is used by the National Cancer Institute for exploratory spatial data analysis, and the *GeoViz Toolkit* (http://www.geovista.psu.edu/geoviztoolkit/index.html), developed by Hardisty and colleagues at the University of South Carolina for more rapid, user-friendly deployment of visualization components. The *GeoViz Toolkit* re-uses *Studio* components, but does not ask users to decide how to connect them together. Instead, it uses one of the coordinators discussed in Sect. 10.3.4 to connect automatically the appropriate pieces in a pre-defined way. This represents a sacrifice of flexibility (the user cannot override how components connect and communicate with each other) in exchange for ease of development. By making this trade-off, the user can more easily get started with an analysis or exploration task.

10.5.2 Creating Applications and Web Services

The trend towards Web services has begun to have an impact on geospatial processing. As a result, there is now a need to create Web services automatically from *Studio* designs, as well as more traditional applications. The *Studio* developers will be tackling this problem in the near future by developing Web service wrappers for the applications and applets that *Studio* generates.

10.5.3 Bridges to Other Software Environments and Applications

Much useful software for spatial analysis is now available in *Python* (http://regional-analysislab.org/?n=PySAL) or via the OS statistical package *R* (http://www.r-project.org/). The *Studio* developers are currently in the process of creating bridges to both of these environments so their content may be used directly in *Studio* without the need to recode or duplicate their efforts. Here the *Jython* project (http://www.jython.org/Project/index.html) is heavily relied on to facilitate embedded Python scripting inside of Java programs, and the inter-process communication mechanisms that allow limited forms of connectivity to applications written in older procedural languages.

10.5.4 Automated Building of Designs

As shown above, perhaps the single biggest barrier to the uptake of component-based FOSS4G projects is that the need for developers to have some technical knowledge, albeit significantly less than previously, is not completely removed. To this end, the *Studio* developers have recently invested effort in encapsulating this difficult-to-acquire knowledge in expert systems that, when coupled with automated reasoning, can automatically build workflows for specific, user-initiated tasks (O'Brien and Gahegan 2004; Luo 2007). Such solutions make extensive use of domain semantics and task ontologies, and require flexible reasoning engines. It is hoped that some of these ideas can be released in future versions of *Studio*.

10.6 Conclusion

There is a continuum of ways in which community-based development of OS software products can be enabled, each varying in terms of the prior commitment developers must make to well-defined data models and interfaces in order to connect their work to the collective whole. At one endpoint of this continuum are systems that rely completely on well-defined, public (and sometimes rather complex) interfaces. *GeoTools* (see Chap. 8) is an example of this latter approach, in which interfaces and data models are defined to OGC specifications (http://www.opengeospatial.org/standards), and it is the completeness of these specifications that enables the distributed development community to add in their own content. At the other end of the continuum are systems that reduce or avoid using specific data models and interfaces, with a focus on generic, simple models that developers can easily map their work onto. GeoVISTA *Studio* is closer to this endpoint.

Of the eight design considerations suggested in Sect. 10.2, *Studio* contributes positively to each as follows:

1. *Studio* is readily extensible. Users can add (and have added) new content that they have built or that they have gathered from other open source projects.
2. Users can (and do) share their functionality. As an example, many users, including later graduate students have been able to make use of Studio functionality, without necessarily being developers themselves.

3. Since *Studio* designs can be saved and reloaded, users can remember exactly what they have done in the past.
4. Workflows can also be passed to other researchers for independent validation, either as *Studio* designs or as stand-alone applications.
5. In terms of enabling non-programmers to use complex functionality, *Studio* is only partly successful as noted above: not all complexities are hidden from the user, though many are.
6. *Studio* certainly supports many useful tools in the area of geovisualization and knowledge discovery. These tools have helped form the basis for many research projects and spin-off products, also as noted above.
7. A good deal of code developed over several years by a cadre of graduate students and three different lead developers is still in use today, and in many cases has been enhanced over the years. The developers have not been successful in maintaining it all, but the better engineered and more useful pieces persist.
8. The mechanism for building lightweight applications from *Studio* designs is a very useful way to share functionality (and good for developing simple tools for teaching). However, this mechanism has been the most frustrating aspect of *Studio* over its history, because changing Java specifications have made it difficult to keep this feature working at times. As noted above, it is now time to move to a Web services model for application development.

Bearing in mind the opening quote, we anticipate that the day will come when *Studio* stops being a success, and instead becomes an 'obstructive nuisance'. However, developing components that are specifically designed for reuse, within a flexible environment in which they can be recombined to suit new needs, certainly staves off that inevitability.

References

Bishr Y (1998) Overcoming the semantic and other barriers to GIS interoperability. Int J Geogr Inf Sci 12(4):299–314

Chin G Jr, Leung LR, Schuchardt KL, Gracio DK (2001) Conceptualizing a collaborative problem-solving environment for regional climate modeling. In Proceedings of the 2001 International Conference on Computational Science (ICCS 2001), San Francisco, CA. Springer-Verlag, Berlin

Dai X, Hardisty F (2002) Conditioned and manipulable matrix for visual exploration. Proceedings of the National Conference for Digital Government Research 2002

Demšar U (2006) Investigating visual exploration of geospatial data: an exploratory usability experiment for visual data mining. *Computers, Environment and Urban Systems* (accepted). Short version presented at the 1st ICA Workshop on Geospatial Analysis and Modelling, Vienna, July 2006

Demšar U (2007a) Knowledge discovery in environmental sciences: visual and automatic data mining for radon problems in groundwater. Trans GIS 11(2):255–281

Demšar U (2007b) Combining formal and exploratory methods for evaluation of an exploratory geovisualization application in a low-cost usability experiment. Cartogr Geogr Inf Sci 34(1):29–45

Demšar U, Fotheringham S, Charlton M (2007) Employing a geovisual exploratory post-analysis for interpretation of results of a spatial statistical method. Accepted to *ICA Visualization 'From geovisualization toward geovisual analytics'*, Helsinki, August 2007

Edsall RM (2003) The parallel coordinate plot in action: design and use for geographic visualization. Comput Stat Data Anal 43:605–619

Ellul C, Haklay M (2006) Requirements for topology in 3D GIS. Trans GIS 10(2):157–175

Fisher RA (1936) The use of multiple measurements in taxonomic problems. Ann Eugen 7(2): 179–188. In Fisher RA (1950) *Contributions to Mathematical Statistics*. John Wiley, New York

Gahegan M, Brodaric B (2002) Computational and visual support for geographical knowledge construction: filling in the gaps between exploration and explanation. Proceedings of the Spatial Data Handling 2002

Gahegan M, Pike W (2006) A situated representation of geographical information. Trans GIS 10(5):727–749

Gahegan M, Takatsuka M, Wheeler M, Hardisty F (2002) Introducing GeoVISTA *Studio*: an integrated suite of visualization and computational methods for exploration and knowledge construction in geography. Comput Environ Urban Syst 26:267–292

Goodchild MF (2000) Keynote address, *Conference of the Association of American Geographers*, Pittsburgh, 2000

Griss M (2000) My agent will call your agent, *Software Development Magazine*, www.sdmagazine.com/articles/2000/0002/0002toc.htm, Feb 2000

Guo D, Gahegan M, MacEachren AM (2004) An integrated environment for high-dimensional geographic data mining. Proceedings of the GIScience 2004

Hardisty F (2002) Designing and building usable geovisualization tools, *EuroConference on methods to define geovisualisation contents for users needs*, Albufeira, Portugal, 9–14 March 2002

Luo J (2007) The semantic geospatial problem solving environment: an enabling technology for geographical problem solving under open, heterogeneous environments. PhD thesis available from: http://etda.libraries.psu.edu/theses/approved/WorldWideIndex/ETD-1749/index.html

MacEachren AM, Dai X, Hardisty F, Guo D, Lengerich G (2003) Exploring high-D spaces with multiform matrices and small multiples. Proceedings of the International Symposium on Information Visualization 2003

O'Brien J, Gahegan M (2004) A knowledge framework for representing, manipulating and reasoning with geographic semantics. *International Conference on Spatial Data Handling*, Leicester, 2004, 584–603

Okabe A, Boots B, Sugihara K, Chiu S-N (2000) Concepts and applications of voronoi diagrams (2nd edn) John Wiley, Chichester

Peuquet DJ (1994) It's about time: a conceptual framework for the representation of temporal dynamics in geographic information systems. Ann Assoc Am Geogr 84(3):441–461

Plaisant C (2004) The challenge of information visualization evaluation. Proceedings of the IEEE conference on Advanced Visual Interfaces, AVI'04

Robinson AC, Chen J, Lengerich EJ, Meyer HG, MacEachren AM (2005) Combining usability techniques to design geovisualization tools for epidemiology. Cartogr Geogr Inf Sci 32(4):243–255

Sametinger J (1997) Software engineering with reusable components. Springer, Berlin

Szyperski C (1997) Component software: Beyond object-oriented programming. ACM Press, New York

Takatsuka M (2001) An application of the self-organizing map and interactive 3-D visualization to geospatial data. Proceedings of the Sixth International Conference on Geocomputation, Brisbane, Australia

Takatsuka M, Gahegan M (2002) GeoVISTA *Studio*: a codeless visual programming environment for geoscientific data analysis and visualization. Comput Geosci 28:1131–1144

Whitehead AN (1933) Adventures of ideas. New American, New York

Wills AC (2001) Components and connectors: catalysis techniques for designing component infrastructures. In: Heinenman GT, Councill WT (eds) Component-Based Software Engineering: putting the pieces together, Addison-Wesley, New York, 307–319

Chapter 11
Design and Implementation of a Map-Centred Synchronous Collaboration Tool Using Open Source Components: The MapChat Project

G. Brent Hall and Michael G. Leahy

Abstract This chapter discusses the design and implementation of a free and open source for geospatial (FOSS4G) project that takes several existing open source projects and builds an application from them to support various modes of participatory input into spatial decision support. The approach uses Internet-based digital maps as the medium of communication between either co-located or dispersed individuals and/or groups to foster collaborative discussion on spatial decision issues. The application domain is first discussed in relation to the principles of the current project. The foundation projects, metalevel architecture and basic components of the 'MapChat' software project are then outlined, followed by a step-through of its functionality. The chapter concludes with a brief discussion of future enhancements.

11.1 Introduction

In the mid-1990s the publication of 'Ground Truth', a collection of essays on the social dimensions of geographic information systems (GIS) and the use of spatial data, edited by John Pickles (Pickles 1995), served to instigate a period of critical reflection that brought about fundamental changes in GIS technologies, the practices of decision making processes in which GIS is used, and the intellectual position of GIScience as a research methodology in geography (Elwood 2006a). Among the issues raised was the concept and nature of participation in GIS use.

In the past decade information and communications technology has expanded to the point where it is now engrained within the daily activities of a substantial proportion of the world's population. Within this milieu the advent and popularity of Internet-based digital mapping with tools such as Google Maps and Google

G. Brent Hall
School of Surveying, University of Otago, Dunedin, New Zealand,
e-mail: brent.hall@otago.ac.nz

Michael G. Leahy
Department of Geography and Environmental Studies, Wilfrid Laurier University, Waterloo, Ontario, Canada, N2L 3C5, e-mail: mgleahy@alumni.uwaterloo.ca

G.B. Hall, M.G. Leahy (eds.), *Open Source Approaches in Spatial Data Handling*.
Advances in Geographic Information Science 2,

Earth have highlighted the uses of spatial data and digital maps. Two important implications of Google Maps are that interaction with maps as media has substantially changed from paper to digital form and, partially because of this, spatial data sources and stores are more accessible to more users than ever before.

The decomplexification or 'dumbing down' of digital mapping and spatial data achieved by Google provides a level of Web-map interaction for the average person that very few, if any, commercial Web mapping tools have been able to achieve. Indeed, communities of developers of Google Maps have emerged to create applications that satisfy their own and other user's online mapping needs (see, for example, http://www.googlemapsmania.blogspot.com). These communities continue to expand with an ever broadening range of applications.

While this achievement, founded on a single product, is impressive, it does not serve as a panacea for the on-going challenges that must be overcome to achieve broad-based participation in the use of GIS technologies. Despite this, Google's characterizations of people, places and their intersecting needs has served to blur the boundaries between traditional GIS applications and user-defined digital cartography, rendering the former less daunting and the latter more appealing. Hence, the most valuable function of the Google-type approach is not complexity, but rather creating the 'capacity [for users] to experiment with spatial representations and produce visual texts to support shifting portrayals of community space and needs' (Elwood 2006b, 704). This, perhaps more than anything else, explains why Google Maps has succeeded where traditional and Web-based GIS has, to date, struggled to achieve broad-based use.

This chapter describes the genesis, composition and application of a FOSS4G project that seeks, like Google Maps, to blur the boundaries between traditional GIS and user-defined map interactions. The project, named MapChat, unites FOSS4G Web mapping with an instant messenger-like synchronous chat interface, using maps and the Internet as the media of communication between users. Geographically dispersed and/or co-located participants can engage in synchronous map-related discussions and strategizing while annotating map features or elements with shared comments, and create new freehand features (points, lines or areas) that can similarly be annotated and discussed by other users.

11.2 Genesis and Principles of MapChat

The MapChat project emerged out of participatory GIScience research in academia, and its basic objective is to achieve broad-based use of digital maps as communications media. MapChat seeks to connect participants to discuss idiosyncratic collaborative perspectives on a common geographic area and spatial decision issue (Jankowski and Nyserges 2001; Balram and Dragicevic 2006).

11.2.1 Co-located and Distributed Participation

Few would question the necessity of soliciting broad-based inputs from multiple sectors of society before arriving at and implementing decisions that will impact upon the public interest. Prior to the digital communications age such input was sought through public meetings, which required the physical presence of individuals and groups at the same place and the same time in order to voice their views. In the last decade, the advent of the Internet in general and Web-based mapping services in particular have allowed participatory inputs into spatial decision problems to be sought from geographically dispersed individuals and groups, without the need for same time, same place presence.

In addition to soliciting inputs from dispersed individuals and groups, MapChat can also satisfy the traditional co-located participatory approaches, where participants work face-to-face on evaluating a decision problem in a common place and time (Jankowski and Nyserges 2001). For dispersed users, all known Web-based participatory GIS approaches and tools are asynchronous in nature. Users interact with spatial data and their manipulations are posted back to a spatially enabled Web server in the form of new map layers or modifications to existing map layers, or some combination of both. Some tools allow participants to interact with other participants' data, but generally this is not the case and there is no direct communication between participants.

There are relatively few tools in existence that allow individuals to interact synchronously to discuss and share perspectives on spatial decision problems. Where synchronous collaboration tools do exist (for example unype.com, which merges Google Earth with a variety of voice and instant messaging tools and social networking applications) they are in their formative stage of development. Synchronous or active collaboration has numerous advantages over asynchronous or passive collaboration in that individuals can join live discussions and share perspectives and map views across the Internet in near to real time. The cornerstones of this conceptualisation of participation in MapChat's basic design are flexibility in use and interactivity between participants. The decision forum is virtual, hence the need to be in a particular place to participate is negated. If participants wish to communicate with each other synchronously they only need be on-line at the same time.

11.2.2 Reusable Component Architecture

To achieve flexibility of use in MapChat, it was important to design and implement procedures for individuals to create and manage groups and to join existing discussions. These functions, which are built around the project's database schema, use code modified from an existing FOSS4G project (Hall and Leahy 2006). This modular use and reuse of generic components is implemented in HTML, JavaScript and PHP so that any other FOSS4G project deploying these languages and an object-relational database schema can include and adapt code directly from MapChat's components.

The approach subscribes to an open system concept in three ways, namely it employs open standards in terms of the data models used and the overall system specification, it uses open source coding in two high-level scripting languages (JavaScript and PHP) that have minimal dependencies, and it uses a flexible architecture to allow the integration of modular tools to meet end user needs. The implications of this open approach are potentially far reaching in that developers can more easily graft components from existing FOSS4G projects to a new project without having to redevelop original code and without having to use a whole project in order to access parts of it. The advantages of this are clear in that they involve less overhead as well as faster potential prototyping of software.

11.2.3 Support for Explicit Data Models

One of the key benefits of the approach taken for the development of MapChat is the variety and flexibility of data models that can be used. The use of MapServer (see Chap. 4 of this text) as the Web-mapping backbone supports a wide variety of spatial data formats with minimal or no data pre-processing required. This is achieved through various methods, including client access to WMS or Web Feature Services (WFS) and native access to common raster and vector data formats. Additionally, MapServer can indirectly access up to approximately one hundred different raster and vector formats through the Geospatial Data Abstraction Library (GDAL) and its accompanying Simple Feature Library (OGR) (see Chap. 5). Further, use of the Cartographic Projections Library (PROJ.4) enables MapServer to merge datasets with different projection systems into a single map that uses one specific projection.

This ability to utilize many different data formats is particularly beneficial for spatial planning and multi-participant decision-making. In these contexts, multiple spatial datasets are usually required for visualizing images or vector features in maps and to serve as reference objects in a discussion. Such datasets often originate from a variety of primary and secondary data sources that use different formats and projection systems. In the case of MapChat, each discussion is based on a single map that is generated from the contents of a MapServer mapfile that identifies source datasets and defines how to display them as layers in a map image. Essentially, through the use of MapServer's programmable objects (e.g., using MapScript in PHP), the complexity of the underlying datasets is hidden by common objects used in the code development.

While the flexibility afforded by MapServer's broad compatibility is important, the support specifically for drawing features dynamically retrieved from PostGIS-enabled PostgreSQL databases is a key aspect of MapChat's development. The use of PostgreSQL as the database management system was a natural choice to support a tool that requires frequent recording and querying of data on-the-fly. In the specific case of MapChat, however, there is the additional requirement for storing, processing and querying spatially-referenced information related to elements of a discussion. In this case, PostgreSQL is used with the PostGIS extension to embed these capabilities seamlessly into the database environment.

Widespread acceptance of PostGIS also means that many other geospatial tools, particularly in the FOSS4G world (e.g., MapServer, Quantum GIS, Kosmo, etc.), are able to use PostGIS-enabled databases as direct input. Thus, by using MapServer's native PostGIS support any new geographic features created during discussions can be displayed simultaneously to each participant. Furthermore, the integration of threaded messages between multiple participants within a single database environment provides a unique opportunity for review and analysis of discussions and their outcomes for planners and decision-makers. Prior to discussing the design and implementation of the MapChat project, two instrumental sets of considerations are noted regarding the project's implementation.

11.2.4 Academia and Project Continuity

A significant general obstacle for the MapChat project is its academic origins. This may seem somewhat paradoxical. However, academic environments, while conducive to creativity and innovation through research, are highly volatile in terms of sustainability and support over the time frame that is required to take an idea from creation through incubation to a practical product. There are several reasons for this, not the least of which is that most research activity is the result of collaboration between professors and graduate students. Master's level programs typically have an eighteen to thirty month duration that includes course work and thesis research and writing. Doctoral programs span three to five years with approximately half of the time spent on the research component. Hence, the amount of time that can be devoted to producing new 'products' is by necessity quite limited.

Combined with the problem of competing demands, most graduate students move on after they complete their degree, hence there is a constant transitional cycle of individuals joining and leaving research teams. This makes the sustained development of complex software difficult relative to the product development life cycle. A solution to this problem is to move promising projects out of academia as soon as possible into a neutral setting where their development can be formalized in a more orderly process (see the geospatial Web concept discussed in Chap. 3). The presence of OS hosting sites such as SourceForge (http://www.sourceforge.net) and the formation of incubator projects by the Open Geospatial Foundation (http://www.osgeo.org/incubation) offer possibilities in this regard. However, they do not necessarily reduce the time and organisational commitments that are required to bring ideas to fruition.

Despite these constraints it is possible to develop successful FOSS4G projects from a University base. For example, the MapServer project has retained its historical University ties, while fostering a broad-based network of project developers. To achieve this level of development requires tight management, a strong institutional commitment, and a dedicated core of University-based and ex-University developers working collaboratively on enhancing the project. The Geo*Vista* project from Penn State University (see Chap. 10) is a further example of a successful University-based FOSS4G project.

11.2.5 Licensing, Intellectual Property and Copyright

The issue of open source licensing is not only a matter of selecting an appropriate license from the relatively large number currently available (see the list at the Open Source Initiative: http://www.opensource.org/licenses/). This is especially relevant in an academic environment, where institutional policies regarding ownership of works undertaken on University-owned equipment and during University-controlled time must be taken into consideration. In collaborative endeavors that involve the work of many contributors issues of intellectual property (IP) and copyright must also be considered. A further factor in academia concerns works that are developed as part of research funded by external agencies, which may lay some claim to ownership if project funds have been expended in their creation of products with commercial potential.

Using the MapChat project and the University of Waterloo's Policy on Intellectual Property Rights (Policy 73 – last updated June 19th, 2000) IP created at the University of Waterloo belongs to the creator. That is, if a student, post-doctoral researcher, or professor helps to produce a 'knowledge-based product' they are automatically a part owner of the IP that is embodied in the product. There are three important caveats to this general conference of rights. First, the University normally retains ownership of IP rights in works created as 'assigned tasks' in the course of administrative activities. Second, owners of IP rights in scholarly works created in the course of teaching and research activities grant the University a non-exclusive, free, irrevocable license to copy and/or use such works in other teaching and research activities, but excluding licensing and distribution to persons or organizations outside the University community. Such licensing and/or distribution activity would be authorized only by an additional license from the owner(s). Third, in sponsored or contract research activities, ownership of IP rights may be determined in whole or in part by the regulations of the sponsor or the terms of the contract. Participants in these research activities must be made aware of any such stipulations of the contract by the Principal Investigator, or the leader of the research project.

While conditions one and two above do not impede in any way the spirit and licensing of an OS project using virtually any of the OS licenses available, the third may be somewhat more problematic for University-sponsored or University-funded research activities. In the case of the funding agency that supports the development of the MapChat project, namely the Geomatics for Informed Decisions (GEOIDE)

Network of Canada's National Centres of Excellence program, IP and copyright issues are referred to policies at the recipient's place of work. Hence, in the case of MapChat, there is no existing or potential conflict between the developers and the licensing of the works undertaken through an OS license. However, this is an important issue that may differ at other institutions depending on their IP and copyright policies.

11.3 Metalevel Architecture and Key Components

This section first discusses the foundation projects that MapChat uses. User and chat project management is then detailed in relation to the database schema discussed in the previous section. The management of chat threading and differentiation between public and private message management are outlined. The section finishes with the extension of chat management into the linkage of chat elements with existing and user-defined map-based features.

11.3.1 Integration of Components with Custom Coding

The MapChat tool relies on several OS projects commonly used for Web development and geospatial applications. As noted above, the main components include PostgreSQL with the PostGIS spatial extension, the MapServer Chameleon template, for designing Web-based mapping applications, server-side scripting using PHP, and the PHP MapScript library that provides MapServer functions within PHP code.

The MapChat tool itself consists of a Web-based portal that provides secure user authentication (login), access to hosted discussions, and management tools programmed in PHP. The MapChat discussions are presented in a Web-mapping interface generated from a Chameleon template. For this interface, a series of customised widgets were designed to provide the unique controls and functions required by the tool. Further, in order to create the synchronous multi-user environment required for MapChat, functions using Asynchronous JavaScript and XML (AJAX) were built into the interface. Upon receipt of new data by the client interface, functions are triggered depending on the content of the data (e.g., a chat message from one user to another in a MapChat discussion), which in turn dynamically update the HTML presented to the user in his/her Web browser.

Figure 11.1 illustrates the general architecture of MapChat based on these components. In general, the tool can be hosted from any PHP-enabled Web server that has local or remote access to a PostgreSQL/PostGIS database. Users can login and use the tool using any modern, standards-compliant browser that supports the use of dynamic HTML (DHTML) and AJAX. The main mapping interface is rendered

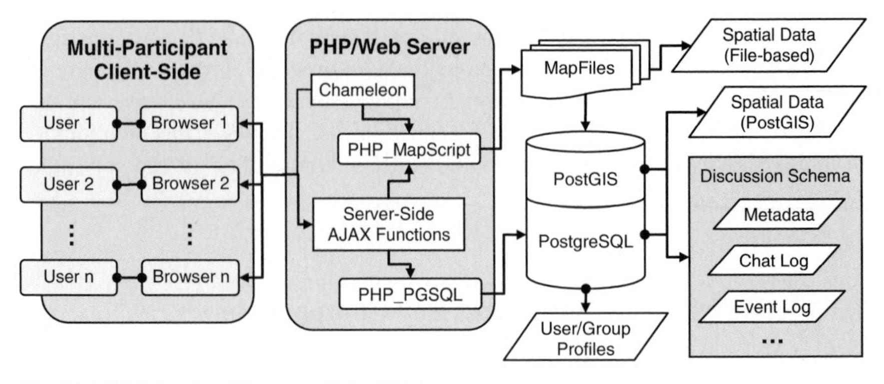

Fig. 11.1 High level architecture of MapChat

from the Chameleon template and sent as HTML to the browser, while individual AJAX scripts receive input and return data in response to specific events called by the browser as users interact with the tool. Event data returned from AJAX requests are then processed by corresponding callback JavaScript functions in the browser. This is illustrated by the general process flow in Fig. 11.2.

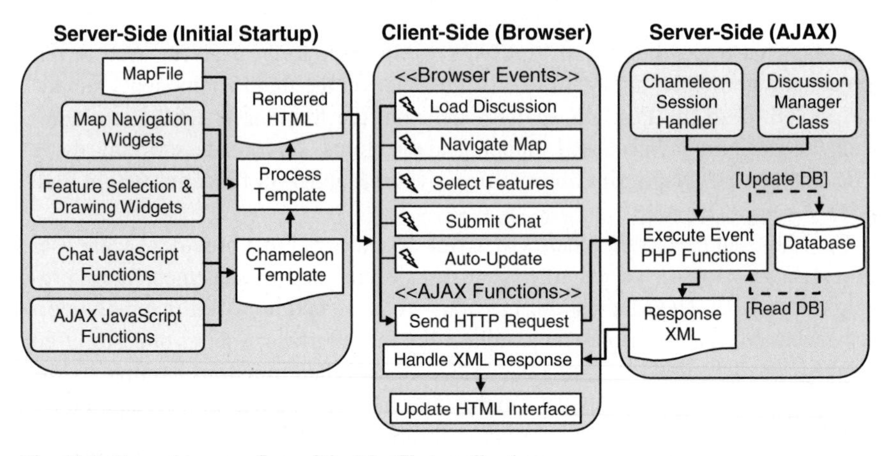

Fig. 11.2 General process flow of the MapChat application

Scripts that were programmed to serve either as program objects that abstract components of the underlying system and concepts, or as 'widgets' that are incorporated into the Chameleon-based interface, underlie the application's logic flow. Table 11.1 summarizes the key components used throughout the operation of the tool. These are discussed in greater detail in the following sections.

Table 11.1 Custom components programmed specifically for MapChat

Component	General Description
Discussion Manager	A class object written in PHP code, which provides various functions for creating, managing, and accessing discussion schemas and their data.
AJAX JavaScript Functions	Functions that are loaded with the HTML in the Web Browser that are specifically designed to submit HTTP requests using the XMLHttpRequest object in JavaScript, and to handle data returned by the server within one or more 'event' elements in the response XML. Each event element may trigger a callback function in the JavaScript code loaded in the Browser.
Database & Discussion Admin Tools	A set of PHP scripts written to allow management of the users, groups, and discussions stored in a MapChat database. They make use of the above Discussion Manager and AJAX components, but operate separately from the MapChat tool itself.
Map & Navigation Widgets	Standard widgets that are written for use in a Chameleon template to create a map interface with standard navigation controls in an HTML document. These have been specifically configured to make use of the AJAX functions after they have been loaded into the Browser, minimizing the execution time and the amount of data transferred.
Feature Drawing and Selection Widgets	A set of widgets based on the same framework as the standard map and navigation widgets noted above, which provide tools for drawing and/or selecting vector features through interaction with the map interface.
Chat JavaScript & PHP Functions	A set of JavaScript and server-side PHP coded routines that are loaded with the MapChat Chameleon template and that provide the functions required to send/receive chat messages to/from the server.

11.3.2 MapChat Discussion Database Structure

Given the nature of communication and interaction that needs to take place between MapChat users involved in a Web-based discussion, an appropriate database schema was needed to facilitate recording and management of user data generated through interaction with the software. Figure 11.3 shows the general structure of the database schema used for MapChat. A set of tables stored in a public schema are used to store general metadata describing discussions, users, groups, and membership information. This basic information is presented to users and managed by an administrator through the main website.

For each discussion that is hosted in MapChat, an individual schema is created in the database to record all of the data generated through users' participation in the discussion. The configuration of the schema used for individual discussions supports the core functionality of MapChat. Two internal metadata tables store lists of member users and groups in a discussion as well as corresponding permissions that may be used to control what types of actions a given user deploys during a discussion.

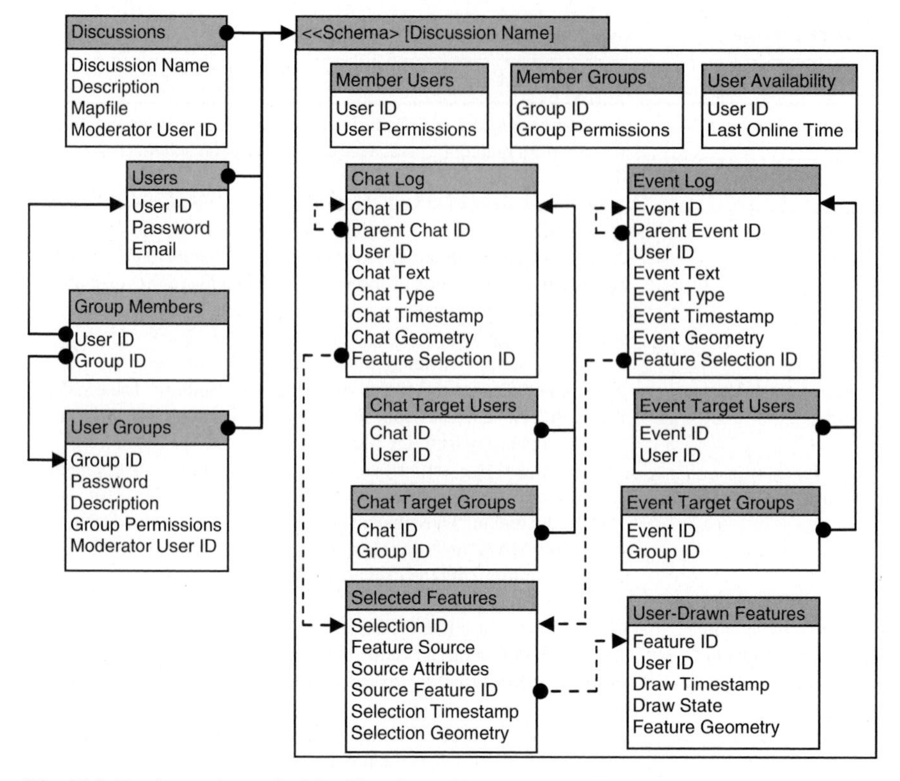

Fig. 11.3 Database schema for MapChat discussions

A third metadata table is used to maintain each user's status information, including how much time passed since the user's last online activity.

The remaining tables and their associations reflect the elements required to capture the progression of a threaded spatially-referenced discussion. Chat messages contributed by participants in a discussion are stored in a chat log table. Each record is assigned an identifier, and where a given message is in reply to another it also has a parent identifier assigned to it. This creates a self-referencing table that can be recursively queried to reconstruct the hierarchical pattern of a threaded discussion. Two additional tables contain records that may be used to identify target users or groups for a given message (e.g., to identify messages sent privately between individual users). Each chat message is also recorded with relevant attributes that identify the contributor, the message text, the type (e.g., to identify explicitly the nature of a given message), and the time of the contribution.

The last two fields of the chat log table provide the spatial dimension to the discussion through the use of PostGIS geometries. The chat geometry field can be used to give a single geographic reference to the chat (e.g., a polygon covering the map extents the user was viewing in MapChat while submitting the message). The second field is used to link to records in a table that lists selected features. The features identified in this table can be derived from existing vector layers in the map, or user-drawn geographic shapes created on the map during the discussion, and are stored in the user-drawn features table.

This approach for containing the discussions within discrete schemas makes the database flexible and extensible. This is exemplified by the additional event log in the schema illustrated in Fig. 11.3. This event log follows the same structure as the chat log, but is specifically used for recording non-chat events related to users' interaction with the tool. This allows not only the discussion and its components to be reconstructed, but also the reconstruction of participant map and data interactions during the discussion. The schema-oriented approach also allows new resources to be added to individual schemas to support specific tools without interrupting the content or structure of the schemas used by any other discussions.

11.3.3 Management of Users, Groups and Discussions

Because of the multi-user nature of the MapChat application, utilities were programmed for the Web portal that specifically manage many users, groups and discussions within the MapChat database. Essentially, these utilities were designed not only to allow flexible control over users' access to utilities and discussions, but to also allow some management responsibilities to be delegated to non-administrative participants.

Initial user registration follows the same procedures typical of many Web-based registration forms, where the user completes a series of form fields though the MapChat portal, then authenticates through a link sent by email from the server to confirm the user's identity. Alternatively, an administrative user can access a form available at the MapChat portal to create user accounts manually, skipping the need for the registration process. This latter approach may be preferable for creating anonymous accounts, for example when conducting case studies where users may be known to each other but wish to remain anonymous to protect the integrity of their views.

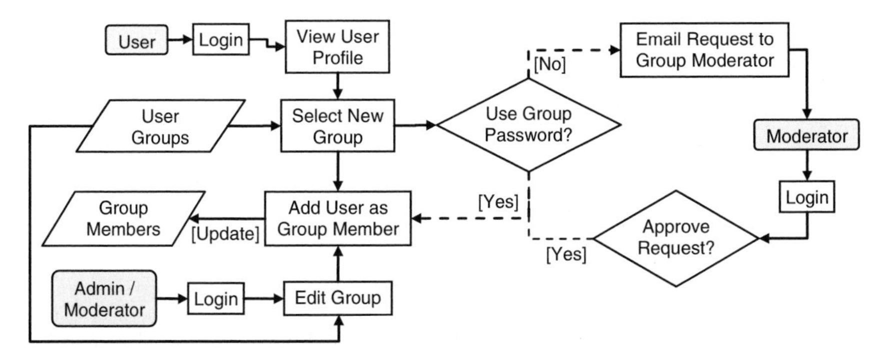

Fig. 11.4 The flow of different methods for registration, group-assignment, and discussion management in MapChat

An individual user account will not have any special privileges on its own when it is initially created. In order to gain additional privileges, the user must be assigned membership to one or more groups. As illustrated in Fig. 11.4, this can be accomplished in several ways including manual assignment by administrators, a user request sent to a group's moderator, or use of a group password that may be distributed by other group members or administrators. Once a user is a member of a given group, then all of the permissions assigned to that group will be applied to the user's account.

The use of user-group memberships not only allows global privileges to be assigned to specific users, but it also allows topic-oriented discussions in MapChat to be arranged for specific groups of users. Both user groups and discussions must be deliberately created and managed by administrators using the utilities provided through the MapChat portal (see Fig. 11.5). For example, groups may be created for 'City Planners' and 'City Residents' respectively, for participation in a spatial planning discussion. When a discussion is created in MapChat, individual users and/or groups are included, and specific discussion-related permissions are assigned to each. Thus, an internal discussion might be created to which only the City Planner group is assigned membership, while another might be created to include both

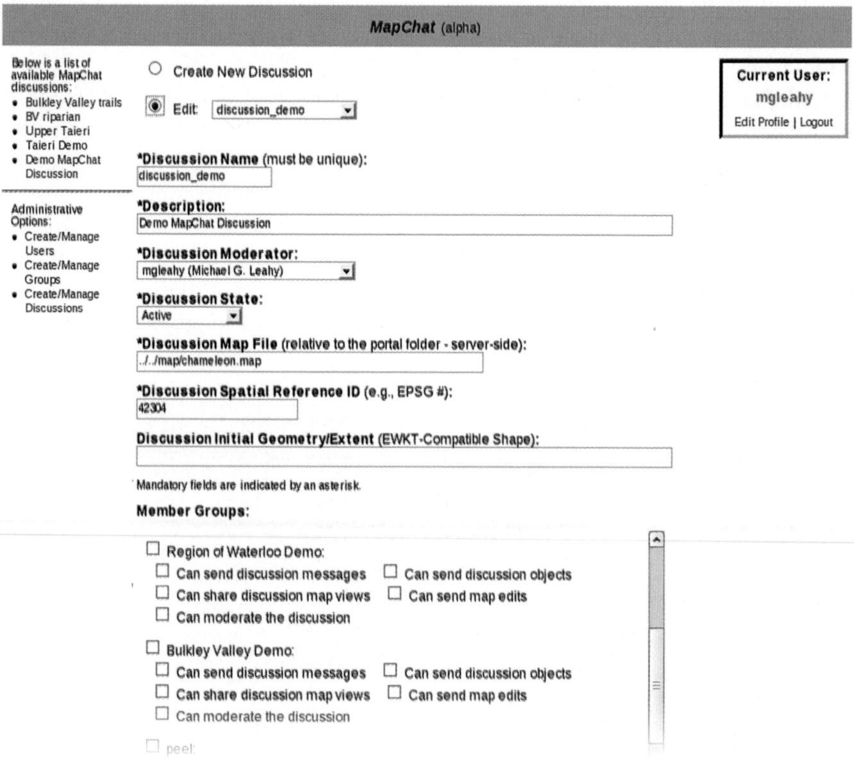

Fig. 11.5 Sample screenshot of discussion properties

groups for a more open public discussion (in which participants could still retain personal anonymity). Alternatively, specific users or groups might be included at different times during the discussion (by editing the discussion's member lists), or some may be included with reduced privileges (e.g., to allow passive observation without options to contribute to a discussion).

In general, the purpose of this approach is to reflect, as closely as possible, the variety of ways that face-to-face public or private meetings and discussions take place, where specific individuals or groups are invited to participate in different ways. Further, the ability to organize discussions according to the decision problem and/or the participating individuals and groups allows for targeted interactions in the MapChat tool and more options for review of discussion outcomes, as described in the following section.

11.3.4 Chat Threading with Public and Private Messaging

The underlying database design and management tools discussed in the previous sections allow discussions for different themes to be organized separately and for chat messages to be threaded. This directly supports the interactive use of the tool during multi-participant discussions. Essentially, each discussion stored in a separate schema can be treated as a general theme for conversation related to the content of a corresponding map. Meanwhile, individual chat messages that start a new thread can be considered to be new topics that can be discussed by subsequent replies, similar to common group email lists or Web-based discussion forums.

Figure 11.6 illustrates the normal sequence of events that take place when a user submits a chat message in MapChat. Initially, the browser submits the message text along with any relevant parameters to a PHP script on the server via a standard HTTP request using AJAX functions. When the data are received, the message

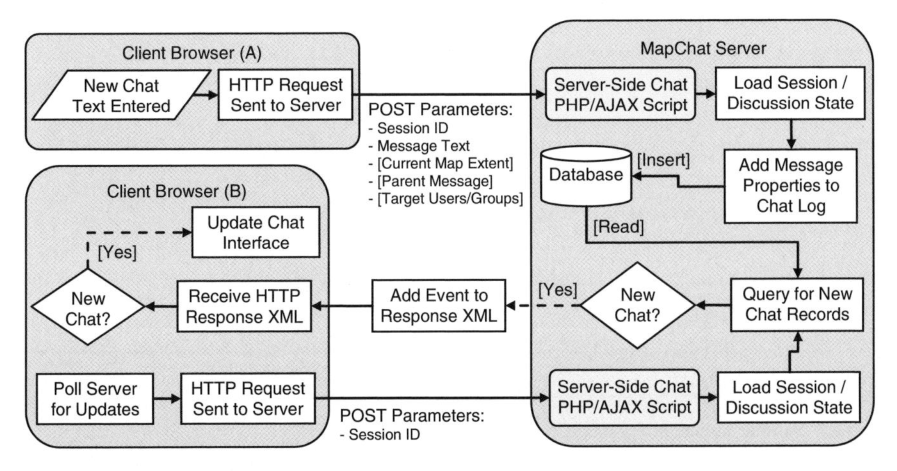

Fig. 11.6 Flow diagram for sending/receiving chat messages

text is inserted into the chat log for the corresponding discussion. If the message is a reply to a previous message, this lineage is recorded along with the message. Any additional data included with the message are recorded if appropriate (e.g., the sender's current map extent, currently selected map features, or the names of recipient users or groups).

Once a message is recorded by the server in the discussion database, it is accessible to other users participating in the discussion. As each user's browser polls the server for updates (currently every five seconds, although this can be adjusted depending on the likely load requirements for a discussion), any new chat messages and their associated attributes that are detected in the discussion's chat log are returned to the client browsers by the server as a response formatted in XML. The AJAX functions in the browser subsequently process the response XML, and update the MapChat chat interface accordingly in the browser. In order to present the threaded structure of a discussion, replies to previous messages are inserted into the chat interface as items in a collapsible tree, formatted akin to a familiar graphical structure used for navigating hierarchically organized data (Fig. 11.7).

Typically, all chat messages will appear for all users participating in a discussion. However, if specific recipient users and/or groups are explicitly identified by the sender of a message, then these messages will only be delivered to the corresponding individuals and/or groups, and will appear in a separate dialog as a private message.

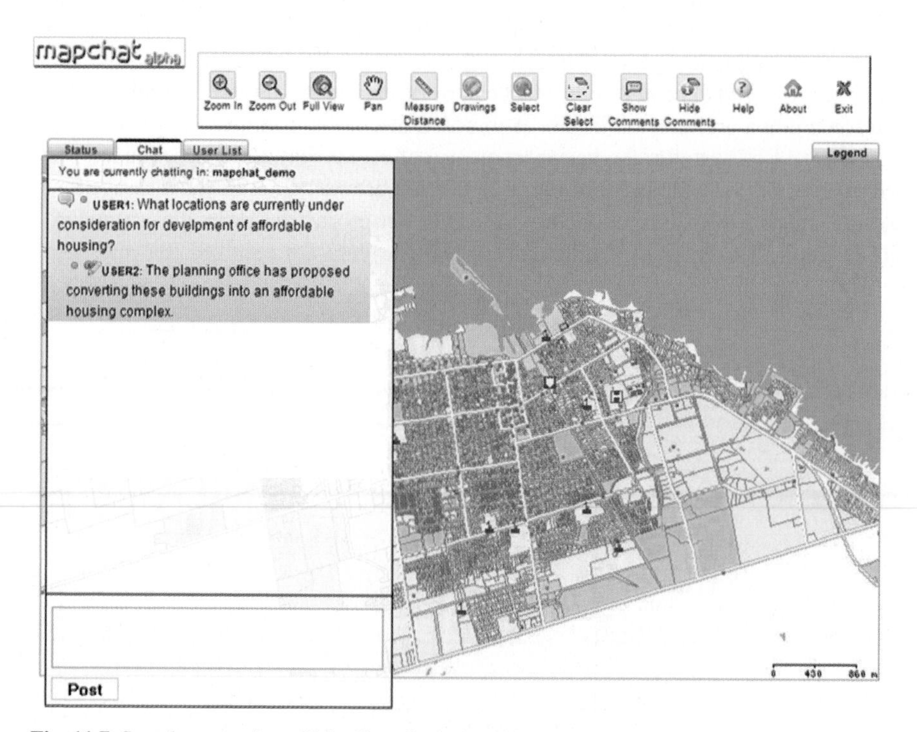

Fig. 11.7 Sample screenshot of MapChat displaying the chat dialog

This functionality is consistent with most other forms of group discussion, where the capacity for private discussion between participants is normally an option (e.g., private messaging in Web-based forums, or emailing individuals registered in an email list). In some contexts, the capacity to do this may be important, where some participants may feel less confident about speaking openly without some level of confirmation from other individuals (similar to individuals who may feel reserved about speaking openly in face-to-face public meetings). In instances where factions exist or evolve in a discussion, the ability to chat and strategize in private prior to re-entering the public discussion is also very important.

Hence, the two underlying design principles of flexibility in implementation and interactivity between participants are both explicitly recognized in the way that the tool manages discussions and allows participants to operate both publically and privately with maximum ease. In addition to reviewing map layers and chatting with other participants, one of the most innovative aspects of MapChat is the ability to link messages explicity with existing map features as well as user defined points, lines or areas that describe features of interest. These linkages persevere in the MapChat map views and database unless explicitly deleted, making them available for individual and/or group review as well as analytic roll back and analysis of discussions relative to the issue being addressed.

11.3.5 Spatial Object Annotation and Chat Linking

The ability of MapChat to provide interactivity to select or draw features on new map layers that are specific to each participating individual and/or group distinguishes it from other Web-mapping and discussion tools. Moreover, these actions are integrated with specific chat messages in the overall discussion. Figure 11.8 illustrates the process of selecting features from existing layers in the map view for a given discussion, and linking them to chat messages that refer specifically to the feature(s) of interest.

When a participant selects features from a vector layer in the map using MapChat's feature selection tool, a record for each feature is added to a table that contains all features selected by users in a given discussion. This table contains both the object itself as a PostGIS geometry as well as any corresponding attributes. This ensures that any information about the feature that may be relevant to the discussion can be reviewed at a later time without referring to the source of the data, while the geometry can be used in spatial queries that use PostGIS functions. As the participants select more objects, additional records are added to the table for each feature, using a common identifier that refers to the current selection set.

A similar process takes place when a user creates a new feature in the map using one of MapChat's four freehand drawing tools (point, polygon, straight line, or irregular line). The new shape's coordinates are recorded as a PostGIS geometry in a table that contains all user-drawn features for a given discussion, and a corresponding record is added to the selection set that refers to the user-drawn feature, as described above.

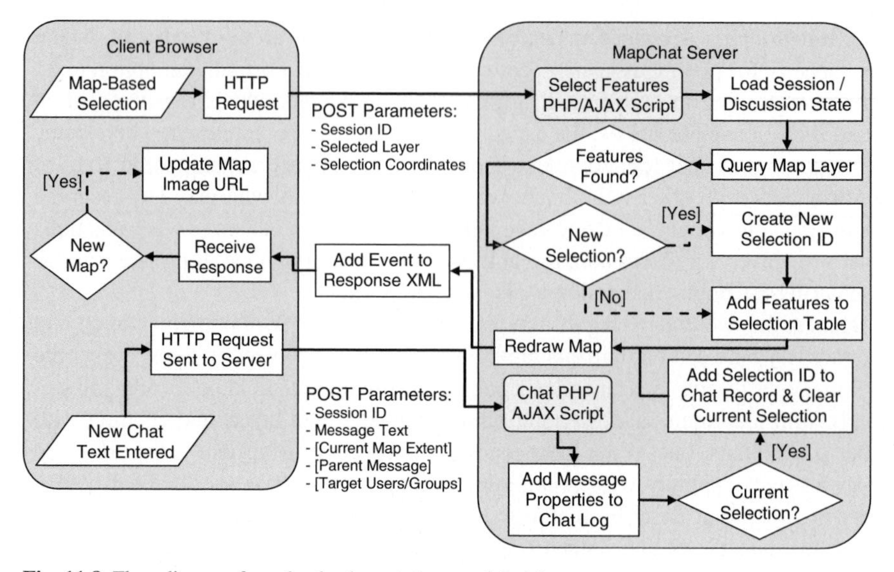

Fig. 11.8 Flow diagram for selecting/annotating spatial objects

If at any time a user submits a chat message while having a set of one or more features currently selected, then the identifier for the user's selection set is included as an attribute for that message. The 'draw status' attribute for any user-drawn features is also set to true at this stage, which causes them to appear as features in the map for all users participating in the discussion. If the chat is private then only individuals participating in the private chat will see the features and their associated comments.

When a chat message has an associated set of features in the map, different ways of exploring the discussion are enabled. In the standard chat window of the user interface, messages that are linked to spatial objects are accompanied by a linked pencil icon. Any user in the discussion can click on this icon to have their map zoom to the extent of the associated features in the map view, and display associated discussion threads inside a Google Maps-like pop-up bubble next to the feature (see Sect. 11.4.4).

Alternatively, discussion threads can be accessed via the features in the map view by clicking the 'show comments' button in the user interface, which displays an icon on top of each feature that is associated with a specific chat thread. Clicking on any of these icons reveals a similar pop-up bubble that displays the related message(s). In addition to enabling more interactive ways of exploring a discussion, this spatial-referencing of chat messages (or annotation of spatial features, known popularly as 'geo-tagging') also allows more in-depth review of comments and map interactions after a discussion has taken place. With the timestamp of each contribution that is stored with all records in the discussion schema, the discussion process can essentially be analyzed in terms of its progression through both time and space, as well in terms of the semantics or arguments used by participants in forming points of view throughout the discussion.

11.4 Low Level Programming Interfaces and Functions

This section extends the discussion in Sect. 11.3 by explaining the lower level programming and interfacing of the project. Object-chat linking, which was discussed in high-level terms in the previous section, is explained in detail in the final section.

11.4.1 AJAX Design

One of the central features of MapChat is its ability to provide an interactive interface and maintain synchronicity between multiple participants that are accessing a server from different browsers without using any additional browser plugins (e.g., Flash, Java, ActiveX objects, etc.). By using AJAX, a standard browser is able to communicate with the MapChat server using the same methods as a standard HTML client, except that the data are sent and received using the XMLHttpRequest in JavaScript code within the browser. The server treats the request and responds as it would to any similar HTTP connection. When the JavaScript object receives the server's response, the data returned by the server are accessible either as normal text, or as an XML document object model (DOM) object (if the response is formatted appropriately by the server).

Using this approach, it is possible to communicate with the server without interrupting the current display presented to a user in the browser window. Were this otherwise, a normal web page would need to be reloaded in its entirety by the browser in order to display new content, or use embedded frames to load new content as separate document objects. Using AJAX, only relevant data need to be sent and received, while small changes are made appropriately to the content displayed by the browser. For example, when a user using MapChat pans or zooms within the map window, a function is triggered that displays a 'wait' image to the user and sends the new map extents requested by the user to the server through an XMLHttpRequest object. The server uses this information to generate a new map image, and then sends a response to the browser with information about the map display (i.e., the actual extents of the new image, the scale, and the Web-accessible location of the map image). When the browser receives this response, it uses it to update any JavaScript or HTML form variables that are relevant to the map's extents or its scale, and it updates the map image in the display with the URL to the new image produced by the server. Once the image is loaded the 'wait' image is removed to notify the user that the process is complete.

While this approach reduces the data processing and transfer time required to update the information displayed in the browser, it has also been adapted in MapChat to act as a surrogate for real-time peer-to-peer connections between users and the server. Rather than using unique functions for handling responses to specific AJAX requests, a common 'dispatch' function is used to handle XML-formatted responses from the server for all events. Each response from the MapChat server is organized as a series of events, where the response data to a given event (e.g., pan or zoom)

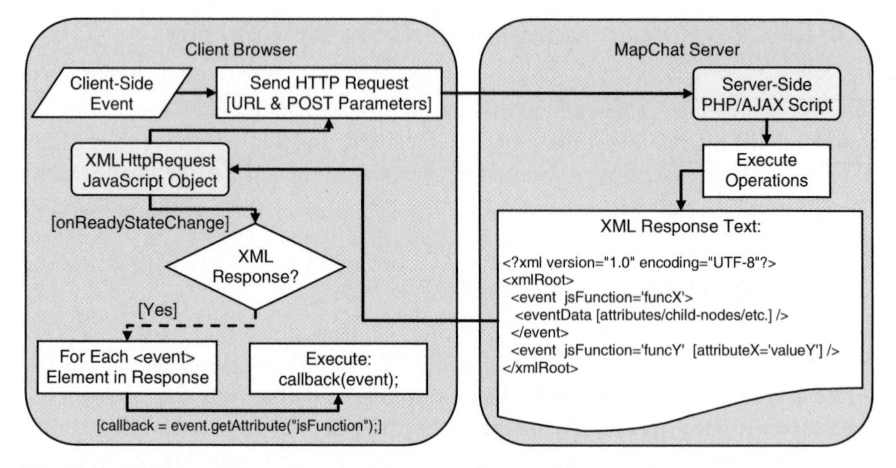

Fig. 11.9 AJAX-based dispatch method for responding to arbitrary events

are contained within an XML element named ‘event’, which is given an attribute named ‘jsFunction’. The value of the ‘jsFunction’ attribute is the name of an appropriate function in JavaScript that corresponds to that particular event. The dispatch function loops through all events contained in a given server response, and calls the corresponding functions named by the ‘jsFunction’ attribute, with the ‘event’ element and the data it contains supplied as an argument. Figure 11.9 illustrates how this takes place.

Using this method, it is possible for the MapChat server to append events in an AJAX response, without the recipient browser anticipating those events. One particular example is after a user zooms in or out of a map view, scale-dependent layers may be revealed or hidden when the map image is updated. If this happens, then the legend must also be updated to reflect changes in the contents of the map view. To ensure this takes place, when the properties of the legend change an additional event element is appended to the response XML after the initial zoom-related event. This event contains the additional required changes to the HTML displayed in the legend, and names the appropriate function for handling these changes. The same method is used effectively to ‘push’ new information to the browser (as a live peer-to-peer connection). For example, if one user has submitted new chat messages to the current discussion, the next AJAX request submitted by other users will contain an additional element in the XML response. That element contains updates to the chat interface containing the new text, and names the appropriate function to process the updates in the ‘jsFunction’ attribute.

11.4.2 User Interface Design and Coding

As noted earlier, the user interface for the MapChat application is built primarily based on the MapServer Chameleon template framework. Specifically, the

Chameleon framework is utilized to process a series of 'widgets' that are called from specialized tags with configurable attributes placed in an HTML-based document that acts as a template. Within PHP code, Chameleon widgets are based on a generic 'widget' class, which provides the basic set of functions and properties required for the Chameleon system. Chameleon itself provides the functions required for processing templates and widgets and for maintaining a user's map and session in PHP.

Each specific Chameleon widget (e.g., the map display and map navigation buttons) is an extension of this generic class. These widgets redefine the functions of the generic widget class to alter the content of the HTML and JavaScript code that is produced as they are rendered from the Chameleon template. The process for each widget involves four general stages:

1) Initialization of the class to read the attributes applied to it in the template;
2) Parsing any URL parameters (sent via subsequent HTTP requests);
3) Defining HTML form parameters, JavaScript variables, JavaScript functions for specific stages of execution (e.g., OnLoad, OnMouseMove, etc.), and/or additional scripts to be called by the browser when the page is loaded;
4) Returning HTML code to be substituted in place of its tag within the HTML template.

When all of the widgets have been processed by Chameleon, the result is a single document that the server returns to the browser, containing all of the appropriate HTML and JavaScript code. This framework makes the process of extending an interface relatively simple, as each additional tool or feature is developed as a separate object that can be added or removed from the template as needed.

For the purpose of MapChat, however, the typical widget design is only used for generating the initial HTML and JavaScript for the interface. Without modification, most of the sample widgets included with the Chameleon software require that the entire template is rendered every time a user interacts with the interface, reproducing all of the HTML and JavaScript, with small updates to the HTML or JavaScript variables. This can become a time-consuming process as many additional, and potentially complex widgets are incorporated into an interface.

In order to reduce the processing required during the use of MapChat, each widget is also accompanied by a small PHP-based script that responds to specific AJAX requests, and returns appropriately formatted XML based on the event dispatching method described in the previous section. This AJAX script only relies on the Chameleon session code to load the user's current map and PHP session in order to process the request and produce the response XML. Thus the Chameleon template and all of the widgets it contains only need to be processed once when a user loads the MapChat interface.

In addition to reducing the processing involved in individual requests when interacting with the interface, it is possible to minimize the initial load time of the interface by deferring some JavaScript code from being loaded until it is actually required. For example, some MapChat widgets use a common JavaScript resource for creating 'pop-in' windows. If these widgets are never used in a given session, the code related to them is unnecessary. Thus, rather than including the code by default when the page is initially loaded, it can be retrieved from the server when it is first requested by a widget. When the code is returned in an AJAX response, it is inserted in a new script element that is dynamically added to the document currently displayed in the browser, making the required functions accessible.

11.4.3 Component Interfaces

As explained in Sect. 11.3 of this chapter, MapChat relies on several underlying components. At the lowest level, these include PHP for the general programming environment, MapServer for interfacing with spatial datasets to generate map images, and PostgreSQL/PostGIS for managing the discussion database. Control of the latter two components is achieved within the PHP scripting language through two object libraries, namely PHP MapScript, which makes MapServer's MapScript functions and classes available in PHP code, and the PGSQL module for PHP, which provides a set of functions and classes for connecting to and querying PostgreSQL databases.

In PHP code, wrapper functions and classes may be used to abstract the most complex aspects involved in using the MapScript and PGSQL modules, providing more generic and useful functions. This is similar to the approach used to create a small set of generic functions for using the XMLHttpRequest object to perform AJAX requests from JavaScript code in the browser described in the previous section. Figure 11.10 illustrates how this is generally accomplished in MapChat, with three general classes (highlighted in grey) providing the basis upon which most of the MapChat widgets and server-side AJAX scripts operate.

The Discussion wrapper class, created specifically for MapChat, provides the basic functions for interacting with MapChat discussions stored in the PostgreSQL

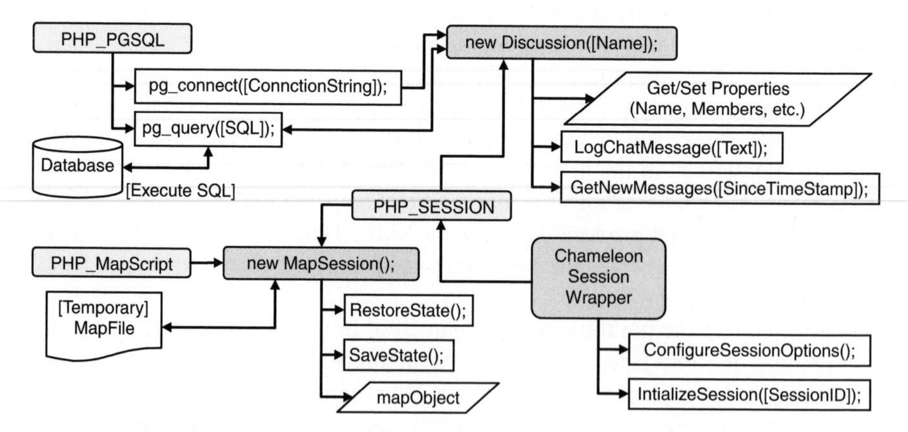

Fig. 11.10 Conceptual arrangement of modules and wrappers used by MapChat

database using the PGSQL module. Using this class reduces the need to interact directly with the schemas and tables in the database with SQL commands, as common operations (e.g., querying for new chat messages) can be abstracted through pre-programmed functions. Another wrapper that MapChat depends on is the MapSession class object that is included with Chameleon. This class provides basic input/output and state management for mapfiles using the MapScript object library.

Finally, the Chameleon session wrapper script is needed in order to maintain a user's current state as subsequent requests are sent from the browser to the server. The user's session is first initialized when the Chameleon template is processed. After this has taken place, when a user performs some action in the interface that submits an AJAX request to the server, the user's Session ID that is sent with the request is used to restore information about the user's session within the PHP execution environment. When the session is loaded, the MapSession and Discussion classes are also restored as needed using parameters saved in the user's current PHP session profile. This allows scripts to interact with the user's current map and to submit or retrieve information from the discussion database.

11.4.4 Object-chat Linking Module Anatomy

To clarify how all of the components discussed in this chapter work together, this section provides a specific example of the sequence of events invoked by the functions executed when a MapChat user selects one or more objects in the map view, and a chat message becomes linked to those objects. The first stage, shown in Fig. 11.11, describes loading of the MapChat interface (refer to Fig. 11.2), at which time the HTML and JavaScript code for the map selection widget is rendered along with the rest of the interface template.

After the user is presented with the interface in the browser, clicking on the map selection button in the MapChat toolbar presents a dialog prompting the user to choose a layer to select features from the interface shown in Fig. 11.12.

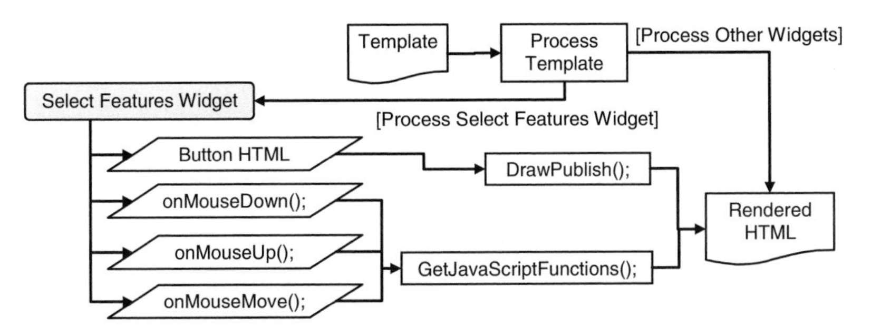

Fig. 11.11 Initialization of the map selection widget in MapChat

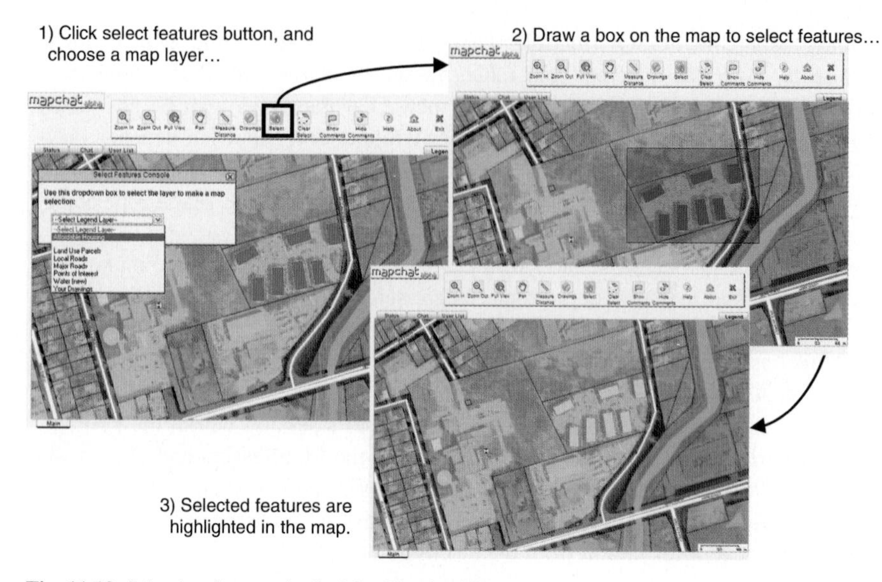

Fig. 11.12 Selecting features in the MapChat interface

Once a layer is chosen, the user must click and drag the mouse on the map canvas to draw a rectangle. When the mouse is released, the map selection tool executes a function that submits the coordinates along with the chosen layer, and the user's session ID to an AJAX script designed for the map selection widget. This script uses the wrappers described in the previous section to load the user's current map view as a MapScript object and queries the selected layer to find the features within the coordinates submitted by the user. If any features are returned, their properties are recorded in the 'selected features' table in the corresponding schema of the discussion database, and a layer is added to the map to highlight these features in the map. If the features are a new selection, then a unique identifier is created to identify the records that are added to the selected features table, while an existing identifier is used if the features are added to a current selection set. This identifier is saved in the user's PHP session, so that subsequent actions can retrieve the corresponding records from the selected features table. Finally, when the AJAX function is complete, a new map view is generated and a response event is returned to the browser that triggers a function to update the display (see Fig. 11.8).

Linking of features to a chat message takes place if the user submits a message using the chat controls while a set of features remain selected in the map. What takes place on the client side is no different than sending a chat message. However, when the server receives the message, additional information is updated in the discussion database. Specifically, when the chat message is received by the server, if the user's PHP session data references a current selection set, then the identifier of that selection set is included as an attribute of the chat message that is recorded in the chat log. Finally, a response event is sent to the user's browser containing the URL to a new map image. The spatially-linked message is sent to other online users, and

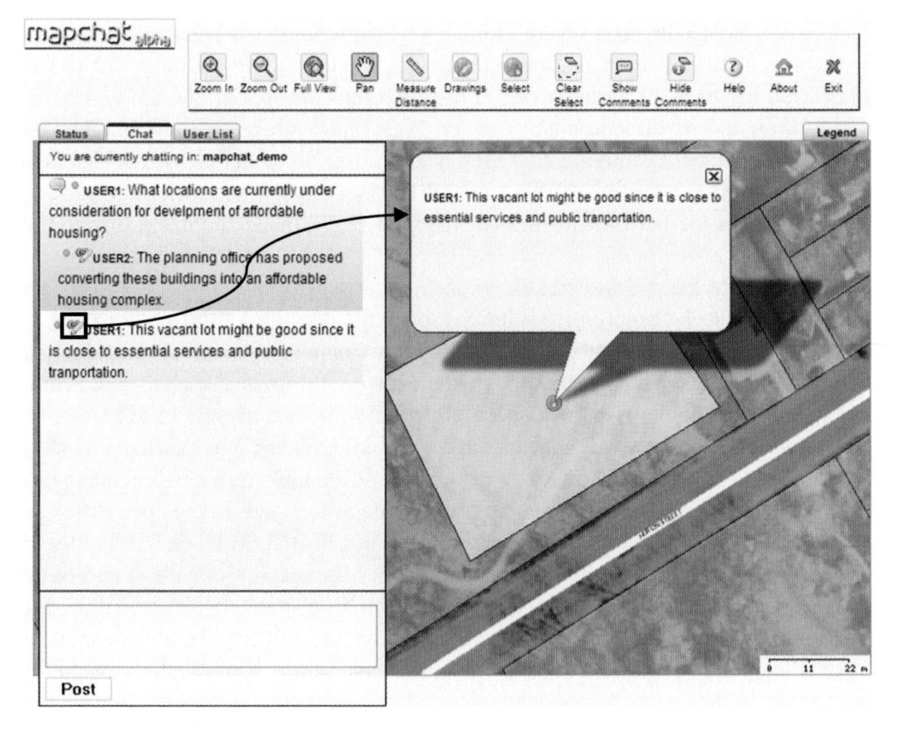

Fig. 11.13 Submitting chat messages linked to selected objects in a map

its selection identifier adds an icon to the message when it is displayed in the chat dialog indicating it is associated with feature in the map. When users click this icon their map view is zoomed to the extent of the associated features with clickable icons superimposed on the map view. When clicked, each icon opens a Google Maps-like chat ‘bubble’ that displays all messages associated with that feature (Fig. 11.13).

11.5 Future Extensions

The first version of MapChat has achieved all of the initial design goals. However, there are several aspects of the software and the broader project that have potential for future development. These are now briefly discussed.

Although MapChat has sufficient functionality to host synchronous multi-user discussions on a spatial planning problem, it lacks tools for users to be able to map their own and others’ discussions to form a visual portrayal of points of consensus and difference in views on the problem at hand. To remedy this deficiency and make the tool more utilitarian for MapChat users and decision analysts alike, a suite of functions are planned to assess argumentation dialogue generated during a multi-participant chat session (Gilbert 1997; van Eemeren and Grootendorst 2004).

Such assessment can take several forms of representation, as discussed by van Eemeren et al. (2002) and van Eemeren and Grootendorst (2004). Rinner (2006) has transferred the basis of earlier work in this area into the spatial domain and Arias et al. (2000) and MacEachren and Brewer (2004) have suggested how spatial data presented via visual displays can be used in a multi-participant discussion. The task remains to program aspects of this research into an OS project such as MapChat. This is high on the list of priorities for the next phase of development during the coming year.

In addition to the need to be able to portray consensus and divergence in points of view and to seek compromise solutions to spatial decision problems, the need to be able to identify a common spatial ontology or frame of conceptual reference for diverse participants in a participatory spatial decision forum is also an area of future application development. In this context, the ability for an analyst to be able to mine a chat database for common concepts and terminologies used by participants may reveal insights into spatial thought processes and reveal how individuals conceptualize space and work through a decision problem cognitively, drawing upon different spatial concepts as they formulate a perspective on a spatial problem. The ability to identify differences and similarities in approaches to spatial recognition and basic analysis may provide useful insights into understanding spatial decision making for diverse groups of decision participants. In MapChat the ability to be able to analyze the chat logs and spatial feature selections and annotations is particularly useful in this regard as these can be rewound, stopped and investigated at any point of the discussion process.

A final area for future exploration with the MapChat project concerns a recurrent theme through many FOSS4G projects, namely how to create greater synergy between projects to relieve the need to recode functions that have already been coded by other developers working on different but related projects. This issue opens the question of the need for improved standards development and their widespread adoption within the FOSS4G community (see Chap. 1). In particular, modularizing functions so that they can be more efficiently hot swapped with the need for only minor revision between projects is something the MapChat developers have experimented with. However, a great deal of work remains to be done in this area.

Acknowledgments We would like to acknowledge the inputs of Dr. Rob Feick, David Findlay, Taylor Nicholls, John Taranu and Brad Noble to the MapChat project. We also gratefully acknowledge the financial support of the Geomatics for Informed Decisions (GEOIDE) Canadian Networks of Centres of Excellence.

References

Arias E, Eden H, Fischer G, Gorman A, Scharff E (2000) Transcending the individual human mind – creating shared understanding through collaborative design. ACM Trans Hum Comput Interact (TOCHI) 7:84–113

Balram S, Dragicevic S (eds) (2006) Collaborative geographic information systems. Idea Group Publishing, Hershey

Elwood S (2006a) Critical issues in participatory GIS: deconstructions, reconstructions, and new research directions. Trans GIS 10(5):693–708

Elwood S (2006b) Beyond cooperation or resistance: Urban spatial politics, community organizations, and GIS-based spatial narratives. Ann Ass Am Geogr 96:323–341

Gilbert MA (1997) Coalescent argumentation. Lawrence Erlbaum Associates, Mahwah, New Jersey

Hall GB, Leahy MG (2006) Internet-based spatial decision support using Open Source tools. In: Balram S, Dragicevic S (eds) Collaborative geographic information systems, Idea Group Publishing, Hershey, pp 237–262

Jankowski P, Nyserges T (2001) Geographic information systems for group decision making towards a participatory geographic information science. Taylor and Francis, London

Pickles J (ed) (1995) Ground truth: the social implications of geographic information systems. Guildford, New York

MacEachren AM, Brewer I (2004) Cartography and GIS: extending collaborative tools to support virtual teams. Prog Hum Geogr 25:431–444

Rinner C (2006) Argumentative mapping in collaborative spatial decision making. In Balram S, Dragicevic S (eds) *Collaborative geographic information systems*, Idea Group Publishing, Hershey, pp 85–102

van Eemeren FH, Grootendorst R, Henkemans FS (2002) Argumentatio: analysis, evaluation and presentation. Lawrence Erlbaum Associates, Mahwah, New Jersey

van Eemeren FH, Grootendorst R (2004) A systematic theory of argumentation: the pragma dialectical approach. Cambridge University Press, New York

Chapter 12
TerraLib: An Open Source GIS Library for Large-Scale Environmental and Socio-Economic Applications

Gilberto Câmara, Lúbia Vinhas, Karine Reis Ferreira, Gilberto Ribeiro de Queiroz, Ricardo Cartaxo Modesto de Souza, Antônio Miguel Vieira Monteiro, Marcelo Tílio de Carvalho, Marco Antonio Casanova and Ubirajara Moura de Freitas

Abstract This chapter describes TerraLib, an open source GIS software library. The design goal for TerraLib is to support large-scale applications using socio-economic and environmental data. TerraLib supports coding of geographical applications using spatial databases, and stores data in different database management systems including MySQL and PostgreSQL. Its vector data model is upwards compliant with Open Geospatial Consortium (OGC) standards. It handles spatio-temporal data

---

Gilberto Câmara
National Institute for Space Research (INPE), Av dos Astronautas 1758, 12227-010, São José dos Campos, Brazil, e-mail: gilberto@dpi.inpe.br

Lúbia Vinhas
National Institute for Space Research (INPE), Av dos Astronautas 1758, 12227-010, São José dos Campos, Brazil, e-mail: lubia@dpi.inpe.br

Karine Reis Ferreira
National Institute for Space Research (INPE), Av dos Astronautas 1758, 12227-010, São José dos Campos, Brazil, e-mail: karine@dpi.inpe.br

Gilberto Ribeiro de Queiroz
National Institute for Space Research (INPE), Av dos Astronautas 1758, 12227-010, São José dos Campos, Brazil, e-mail: gribeiro@dpi.inpe.br

Ricardo Cartaxo Modesto de Souza
National Institute for Space Research (INPE), Av dos Astronautas 1758, 12227-010, São José dos Campos, Brazil, e-mail: cartaxo@dpi.inpe.br

Antônio Miguel Vieira Monteiro
National Institute for Space Research (INPE), Av dos Astronautas 1758, 12227-010, São José dos Campos, Brazil, e-mail: miguel@dpi.inpe.br

Marcelo Tílio de Carvalho
Catholic University of Rio de Janeiro (PUC-RIO), Rua Marquês de São Vicente, 22522. 453-900 Rio de Janeiro/RJ, Brazil, e-mail: tilio@tecgraf.puc-rio.br

Marco Antonio Casanova
Catholic University of Rio de Janeiro (PUC-RIO), Rua Marquês de São Vicente, 22522. 453-900 Rio de Janeiro/RJ, Brazil, e-mail: casanova@tecgraf.puc-rio.br

Ubirajara Moura de Freitas
Space Research and Applications Foundation (FUNCATE), Av. Dr. João Guilhermino, 429 – 18th floor 12210-131 São José dos Campos, SP, Brazil, e-mail: bira@geo.funcate.org.br

G.B. Hall, M.G. Leahy (eds.), *Open Source Approaches in Spatial Data Handling*. Advances in Geographic Information Science 2,

types (events, moving objects, cell spaces, modifiable objects) and allows spatial, temporal, and attribute queries on the database. TerraLib supports dynamic modeling in generalized cell spaces, has a direct runtime link with the R programming language for statistical analysis, and handles large image data sets. The library is developed in C++, and has programming interfaces in Java and Visual Basic. Using TerraLib, the Brazilian National Institute for Space Research (INPE) developed the TerraView open source GIS, which provides functions for data conversion, display, exploratory spatial data analysis, and spatial and non-spatial queries. Another noteworthy application is TerraAmazon, Brazil's national database for monitoring deforestation in the Amazon rainforest, which manages more than 2 million complex polygons and 60 gigabytes of remote sensing images.

12.1 Introduction

Recent advances in spatial databases have changed both the nature and process of geographic information system (GIS) software development. Spatially-enabled database management systems (DBMS) such as PostgreSQL empower a transition from monolithic GIS with hundreds of functions to a generation of spatial information applications tailored to suit specific user needs. These capacities have been a major boon for the free and open source geospatial (FOSS4G) community, many members of which are using the new generation of databases to build unique and innovative applications.

One of the expected impacts of open source software (OSS) is its benefits for developing nations. As Weber (2004) points out, combining OSS with the technical workforce available in developing countries can enable technology transfer. He states, "Of course information technology and open source in particular is not a silver bullet for long–standing development issues; nothing is. But the transformative potential of computing does create new opportunities to make progress on development problems that have been intransigent" (Weber 2004 p. 254).

Following from this point, GIS is a key technology for developing nations in domains such as environmental protection, urban management, agricultural production, deforestation mapping, public health assessment, crime-fighting, and socio-economic measurements. However, the demands of these applications go well beyond the current specifications of the Open Geospatial Consortium (OGC). Large-scale environmental and socio-economic applications compel FOSS4G to include significant spatial analysis capacities to meet the needs of end-users (Goodchild 2003). Hence, FOSS4G should incorporate research advances in areas such as spatio-temporal data models (Erwig and Schneider 2002; Hornsby and Egenhofer 2000), geographical ontologies (Fonseca et al. 2002), spatial statistics and spatial econometrics (Anselin 1999), cellular automata (Couclelis 1997), and environmental modeling (Burrough 1998). These topics have largely been outside the reach of the GIS user community due to a general lack of widely available tools that support them. Incorporation of some of these new techniques into GIS applications is necessary for the user community to extract the full potential of spatial databases.

With this motivation, TerraLib was developed as an open source GIS software library that extends object-relational DBMS technology to support spatio-temporal models, spatial analysis, spatial data mining, and image databases. The design goal for TerraLib is to support large-scale applications using cadastral, socio-economic and environmental data. This goal was a mandate of the main organization that supports TerraLib, the Brazilian National Institute for Space Research (INPE). INPE is Brazil's primary institution for space science and technology. Its mission includes building satellites, developing environmental applications, and producing weather and climate forecasts. Since 1984, INPE has had a research and development division for GIS to support its actions in earth observation and to promote GIS and remote sensing technology in Brazil. The two other main project partners are the Computer Graphics Group (TecGraf) of the Catholic University of Rio de Janeiro (PUC-RIO) and FUNCATE, a non-profit foundation that develops GIS applications using OSS. All organizations involved in the TerraLib project share the same general design goals. Thus, TerraLib is a project with long-term support and a stable and secure working environment for its developers. This chapter describes the TerraLib library, explains the main design decisions, and points out how the library incorporates research results from GIScience in its development.

12.2 Challenges for Innovation in FOSS4G

The OGC specifications noted above provide a sound basis for developing FOSS4G projects. However, many applications need tools which go beyond these specifications. Thus, one of the lines of growth in FOSS4G is to provide new tools for application developers. However, there are pitfalls. Building innovation in open source GIS is a threefold challenge. Given the design goals for the project discussed in this chapter, the first step required selecting, from the large body of GIScience literature, those advances that are relevant to the project's objectives. These advances then need to be implemented in industrial-strength code. The final hurdle is documenting these features and sharing them with the broader FOSS4G development community.

A basic design objective for TerraLib was to support innovative applications to help people and protect the environment. Thus, current GIS research was first evaluated, and ideas and proposals were selected that were relevant to the design goals. This led to concentration in the following three areas:

(a) *Spatial Statistics*: since Anselin's pioneering work on spatial analysis (Anselin 1989), promising advances have appeared in the field of spatial statistics and spatial data mining (Anselin 1995; Fotheringham et al. 2002; Openshaw and Alvanides 2001; Martin 2003). The main focus of these contributions is to improve the ability to extract information for socio-economic data. This is relevant to public policy applications of GIS.

(b) *Spatio-temporal Models*: there are two broad categories of spatio-temporal objects. The first concerns *moving objects*. Moving objects relate to, for example, information about spatial and temporal positions of planes, storms or automobiles. The widespread relevance of location-based applications has motivated developments in the field of moving object databases (Güting and Schneider 2005). There is a large research area in algorithms and query methods for moving objects (Sistla et al. 1997). The second type concerns *evolving objects* that do not move, but whose geometry, topology and properties change. They arise when changes that occur in, for example, cadastral GIS or in land cover patterns are considered (Medak 2001). Evolving objects are important for environmental models, which depict the temporal evolution of a pattern in a landscape. Examples of environmental models include land change models, epidemiological studies, population flows, and ecological mapping (Burrough 1998; Veldkamp and Fresco 1996).

(c) *Remote Sensing, Image Processing, and Image Databases*: remote sensing satellites are the most significant source of new data about our planet, and remote sensing image databases are the fastest growing archives of spatial information. New high resolution optical sensors and polarimetric radars have improved application areas such as environmental monitoring and urban management. There are important recent advances in object-oriented segmentation and classification, and in remote sensing data mining (Blaschke and Hay 2001; Navulur 2006; Aksoy et al. 2004). It is also important to include support for raster data handling in open-source DBMS, following the research results of Chang et al. (1988) and DeWitt et al. (1994).

To translate these ideas to industrial-strength code, the developers of TerraLib first undertook various research projects and published the results from these (Pedrosa et al. 2002; Almeida et al. 2003; Vinhas et al. 2003; Ferreira et al. 2005; Silva et al. 2005; Assunção et al. 2006; Feitosa et al. 2007). These results enabled the TerraLib development team to assess the potential benefits of each technique, as well as the trade-offs needed to generate production code. Software engineering tools for GIS were also examined during this process. One of the conclusions from this was to confirm the usefulness of *design patterns* and *generic programming* as a basis for achieving reuse in GIS software development (Câmara et al. 2001; Vinhas et al. 2002).

The last and most difficult problem is sharing the resulting code with the FOSS4G community. Many of the new tools and techniques might be unfamiliar for FOSS4G developers and practitioners. Hence, there is a need to explain not only the code, but also the ideas behind it. Experience has shown that face-to-face workshops and meetings are the best way to discuss new ideas and their implementation. A second-best alternative is writing detailed documentation, which is not easy to achieve in open source projects (see Chap. 2). Developers have to work hard to share their results, and the TerraLib team is aware of this challenge.

12.3 The Design of TerraLib

This section discusses the requirements and design rationale for TerraLib. It presents the alternatives considered at various points, and explains the final choices that were made. The discussion explains how product requirements led to the software architecture, the conceptual model, and extensions to the basic OGC specifications.

12.3.1 Product Requirements

The main goal for TerraLib led to the following needs:

(a) *Ease of customisation*: developers should require little effort to use the library to develop their applications. They should concentrate only on specific user needs, and the library should provide powerful abstractions that cover the common needs of a GIS application.
(b) *Upward compatibility to the OGC simple feature data model*: considering the impact and popularity of the OGC specifications, a TerraLib spatial database should be compatible with the OGC simple feature specification (SFS). This was not an original project requirement. When the project started in 2002, the developers initially underestimated the impact and extent of the OGC specifications. Hence, TerraLib's code was redesigned (from Version 3.2 to Version 4.0) to satisfy the need for conformance.
(c) *Decoupling applications from the DBMS*: the library should handle different object-relational databases transparently.
(d) *Support for large-scale applications*: to be useful for environmental and socio-economic applications, the library should provide efficient storage and retrieval of hundreds of thousands of spatial objects.
(e) *Extensibility*: a GIS library should be extensible and accessible by other programmers. Introducing new algorithms and tools should not affect already-existing code.
(f) *Enabling spatio-temporal applications*: emerging GIS applications need support for different types of spatio-temporal data, including events, mobile objects, and evolving regions.
(g) *Remote sensing image processing and storage*: the library should be able to handle large image databases, and inclusion of image processing algorithms should be easy.
(h) *Spatial analysis*: there should be support for spatial statistical methods to improve the ability to extract information from socio-economic data.
(i) *Environmental modelling*: there should be support for environmental and urban models, including dynamic models using cellular automata.

To respond to issues (a) and (b), TerraLib has a strong conceptual model, as explained in Sect. 12.3.2. Points (c), (d) and (e) led to a software architecture described in Sect. 12.3.3. The last four issues are considered in Sects. 12.3.4–12.3.7.

12.3.2 Conceptual Model

This section describes TerraLib's conceptual model that was designed to support requirements (a) and (b) noted above. When designing TerraLib, the developers had to make numerous choices which are typical of software library design in general (Meyer 1990; Krueger 1992; Fowler et al. 1995). Apart from basic principles such as applicability, efficiency, ease of use, and ease of maintenance, there are important trade-offs. In this regard, consider two opposing visions:

- *Vision 1*: Libraries should take a minimalist approach. They should provide only primitive building blocks and include generators that can combine these blocks to yield complex custom applications. They should be split into independent modules, with as few dependencies as possible. The developer's focus can be narrowed to those modules that are of interest (Batory et al. 1993).
- *Vision 2*: Libraries should have strong ideas behind them. All the functionalities and modules should work well together. The idea is to maximize reuse by minimizing cognitive distance, which Krueger (1992, p. 136) defines as: "the amount of intellectual effort expended by software developers to take a software system from one stage to another". In this vision, the intellectual effort that software developers need to take a library and development of applications should be minimal. Application programmers use higher-level abstractions to build applications, and do not need to understand the details of the library's source code.

The choice between the two visions depends on a library's initial design goals. For libraries designed to be part of a larger software project, the first vision is the usual choice. Examples include libraries such as the Standard Template Library (STL) in C++ (Austern 1998), or the *shapelib* utility for GIS (Warmerdam 2007 – see Chap. 5). At the other extreme, libraries are designed to be easily extendible to build complete applications. One example includes libraries that use the model-view-controller (MVC) pattern (Krasner and Pope 1988) such as Java Swing (Elliott Eckstein et al. 2002). Libraries with strong concepts dictate how the user should develop the application.

TerraLib follows the second vision, since it aims to make it easy for programmers to develop end-user applications. To do this, the library needs to consider the semantic mismatch between relational databases and object-oriented applications. Relational databases store information in *tuples*, but GIS applications manipulate *objects*. A typical GIS application consists of four steps: (a) query the spatial database; (b) convert the query results (tuples) into objects; (c) manipulate these objects to create new objects; (d) display the resulting objects.

Thus, applications need to distinguish between data sources (the *spatial database*) and data targets (*the set of objects that must be manipulated and displayed*). To reduce the cognitive distance from OSS code to a deliverable application, the GIS developer needs a library that provides abstractions both for the data sources and for the data targets. These abstractions should support the four basic GIS components (query, conversion, manipulation, and display). Consequently, TerraLib supports the following abstractions:

- *Database*: a repository of information that contains data and metadata.
- *Layer*: a container of spatial objects that share a common set of attributes. Examples of layers are thematic maps (soil or vegetation maps), cadastral maps (map of land parcels in a city), or raster data such as satellite imagery. A layer knows its cartographic projection. Layers are inserted in the database by importing data from files or other databases, or by processing other layers. A layer stores the temporal evolution of the objects it contains.
- *Representation*: the geometric parts of data contained in a layer. TerraLib supports different geometries, including two-dimensional (2D) vectors (points, lines or areas), cell spaces, networks, triangulated irregular networks (TINs), and multi-dimensional rasters. The same data can have different representations (for example a city can be represented by the polygon that describes its political boundaries or by a point that represents its geometric centre).
- *Theme*: a theme contains a subset of the objects of a layer, produced by a selection. The selection may use attribute, spatial, or temporal conditions. Each theme has a set of presentation attributes for graphical display.
- *View*: this is a set of themes that are visualized or processed together. It defines a particular user's view of the database. A view has a unique cartographic projection, and the themes it contains are converted to this projection.
- *Visual*: this comprises a set of presentation attributes. Each theme has a unique "visual". A visual includes choroplethic filling and contour colours for polygons, thickness and colours for lines, or symbols for points.

TerraLib distinguishes between data sources and data targets. The abstractions of *database*, *layer*, and *representation* relate to the source domain and describe data organization and hierarchy. The ideas of *theme*, *view*, and *visual* relate to the target domain and describe data retrieval and presentation. A query in TerraLib retrieves tuples from layers, converts these tuples into a set of objects, and groups objects of the same type in themes. Thus, *layer* and *theme* are complementary abstractions. *Layers* organize spatial data in the database. *Themes* organize objects for manipulation and display. Similarly, *databases* and *views* are complementary concepts. A *database* organizes *layers* of spatial data. A *view* organizes *themes* containing spatial objects.

These concepts provide a set of higher-level abstractions on top of the OGC SFS, which are not part of the current OGC model. Terralib stores these entities in a set of metadata tables, built when creating a new database. These metadata tables are kept updated as long as TerraLib manages the database. Should an OGC-compliant application access a TerraLib database, it will only access the tables described in the OGC model.

12.3.3 Software Architecture

This section discusses how TerraLib responds to requirements (c), (d) and (e) as stated in Sect. 12.3.1 (*DBMS-independence*, *efficiency*, and *extensibility*). To address these issues, it was decided to use the C++ programming language for development. The developers had previous experience and had developed many algorithms in C++ as part of SPRING, their earlier GIS project (Câmara et al. 1996). Existing DBMS such as PostgreSQL provide native interfaces in C++. Also, C++ helps with the use of generic programming (Alexandrescu 2001) and design patterns (Gamma et al. 1995).

The developers chose an architectural design that has a kernel and a periphery. Maintenance of the kernel is the responsibility of a core team composed of a few senior programmers. Other contributors use the library's core to add new algorithms that test the library's core for extensibility and robustness. This follows the approach used for successful OSS products such as Linux, PostgreSQL, and Apache, which all have a kernel whose maintenance is the responsibility of a small team. Contributions from the community occur at the external layers. As an example, out of more than 400 developers, the top 15 programmers of the Apache Web server contribute 88% of added lines (Mockus et al. 2002). TerraLib's architecture has four parts, as shown in Fig. 12.1:

- *Kernel*: the core of TerraLib provides a set of spatio-temporal data types, code for cartographic projections and topological spatial operators, an API for storage and retrieval of spatio-temporal objects in databases, and classes for controlling visualization of spatial data.
- *Drivers*: modules that specialize the kernel's generic database application programming interface (API) to allow access to DBMS such as PostgreSQL (with or without the PostGIS spatial extension) or MySQL, and to external files in both

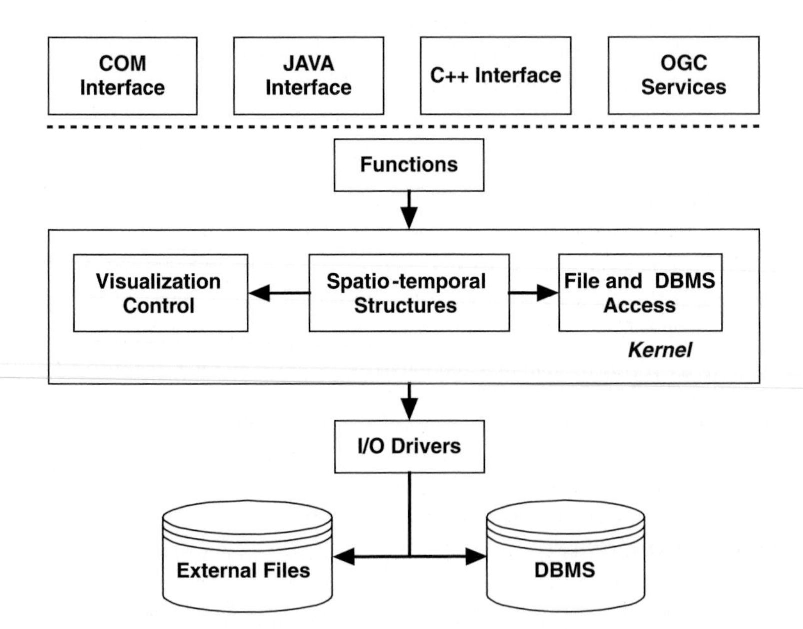

Fig. 12.1 TerraLib software architecture

open and proprietary formats. Basic maintenance and upgrade is the responsibility of the project core team.
- *Functions*: algorithms that use the kernel API. Typical functions include spatial analysis, query, and simulation languages. The functions are designed to allow external contributions.
- *Interfaces*: different interfaces to the TerraLib library that allow software development in different environments (Java, COM) and the support for OGC services such as Web Map Services (WMS), Web feature services (WFS) and Web coverage services (WCS).

The core of TerraLib's kernel is its set of spatio-temporal data types and its methods for query processing. The OGC Simple Feature Geometry model is used for storing basic vector geometries. Extra metadata tables support the abstractions described in Sect. 12.3.2. For a full description of the metadata, see the library's documentation (Vinhas et al. 2007). One key need is efficiency. The developers have spent much effort on issues such as indexing techniques and computational geometry algorithms (Queiroz 2003; Rodrigues et al. 2005; Rodrigues et al. 2006). With this work, the library now supports large-scale geographical databases, as discussed in Sect. 12.4.

The I/O drivers provide the interface between the kernel and the various DBMS and file formats. The library handles different object-relational databases, using a generic database API that handles the specific features of each DBMS. Using this API, a TerraLib programmer can work at an abstract level. TerraLib hides the differences between products such as PostgreSQL/PostGIS and MySQL from the programmer (Ferreira et al. 2002).

As noted earlier, the design of TerraLib *functions* aims for extensibility, as introducing new algorithms and tools should not affect existing code. Adoption of the principles of generic programming and design patterns helped achieve extensibility in this regard (Câmara et al. 2001). Three design patterns were found to be especially useful:

- *Factory*: this pattern provides an interface for creating an object, but lets subclasses decide which class to instantiate. In GIS, it is useful to include new functions without changing existing code. For example, there are hundreds of cartographic projections. When code for a new projection is inserted in TerraLib, it tells the projection factory about its existence and the projection factory calls this new code when it is needed.
- *Strategy*: provides an interface to a family of algorithms, and makes them interchangeable. In GIS, the strategy pattern is useful when there are different ways of performing the same function. This occurs often in image processing. For example, there are many different types of image filters. By using the strategy pattern, a programmer can use the same code for different filters.
- *Iterators*: TerraLib uses iterators to decouple algorithms from data structures. For example, to compute a histogram it is not essential to know if the data are a set of points, a set of polygons, a grid, or an image. The algorithm only needs to look into a list and get the values of the items that satisfy a certain property (for

example, those that are closer in space than a specified distance). In a similar way, most spatial analysis algorithms can be independent of spatial data structures and described only by their properties (Vinhas et al. 2002).

12.3.4 Raster Data Handling

As noted earlier, TerraLib handles raster data types as well as vector data. All raster data types are handled in a unified way, using an interface with two main methods, namely one to set the value of a point on a multidimensional raster and one to recover its value. The library provides decoders for different raster data formats, and iterators for accessing image data and developing image processing algorithms. The decoders are responsible for handling the particularities of each data source. The iterators are specialized pointers that traverse a raster in a predefined way (for example, only inside a given polygon). They hide the internal details of the raster data (Vinhas et al. 2003).

The library has drivers for storing raster data in different DBMS. TerraLib uses indexing and compression to achieve good performance in a standard DBMS, even for large satellite images. Indexing combines tiling and multi-resolution pyramids. Multi-resolution pyramids store the raster data at various sizes and degrees of resolution. Each resolution level is divided into tiles. A tile has a unique bounding box and a unique spatial resolution level, which are used to index it. Figure 12.2 shows a pictorial representation of the tiling and multi-resolution storage model.

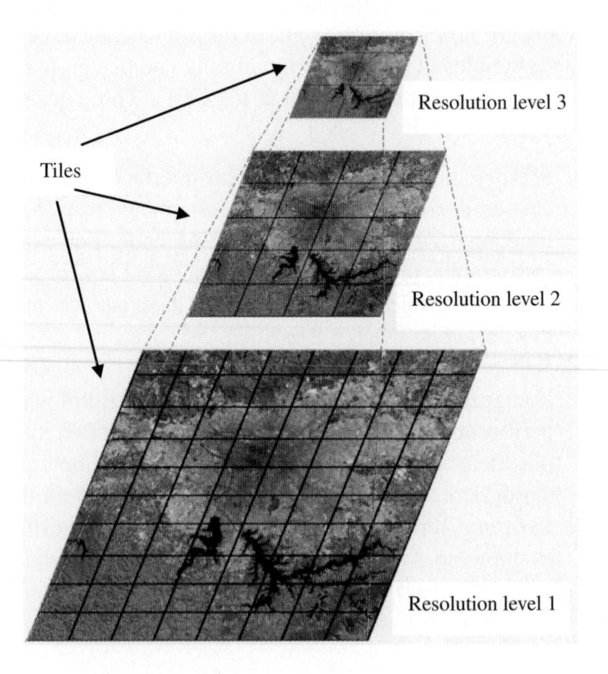

Fig. 12.2 TerraLib raster data indexing

The multi-resolution pyramid approach is useful for display of large data sets, avoiding unnecessary data access. TerraLib stores the whole pyramid. To compensate for the extra storage needs, it applies lossless compression to the individual tiles. When retrieving a section of the data, only the relevant tiles are accessed and decompressed.

TerraLib provides a large set of image processing algorithms including filters, segmentation, classification, mixture models, and geometric transformations. The image processing algorithms have a common interface for receiving as input instances of the raster API and a set of parameters. Two particularly important algorithms are the object-oriented segmentation and region classifier algorithms developed by INPE, originally as part of the SPRING software (Câmara et al. 1996). These algorithms were extensively validated for extracting land use patterns in tropical forests, and were favourably reviewed in a recent survey (Meinel and Neubert 2004).

Creating new image processing algorithms is straightforward. The library has a set of standard protocols that combine the *Factory* and *Strategy* design patterns. First, a programmer develops a new algorithm (for example, a filter), then instructs TerraLib to use the proposed strategy to filter the image. The programming manual provides further details as to the subsequent steps that need to be followed to integrate the new code (Vinhas et al. 2007).

12.3.5 Spatio-temporal Queries

There are numerous different ways to record spatio-temporal information in a database. The main alternatives are to (a) provide snapshots of data, (b) store sequences that describe the temporal evolution of spatial objects, or (c) store both objects and events that change them (Hornsby and Egenhofer 2000; Grenon and Smith 2003; Galton 2004; Worboys 2005). TerraLib adopts a dual perspective, namely archiving fields (stored in raster data) as snapshots and objects (stored in vector data) as sequences. A spatio-temporal object in TerraLib is a sequence of static objects with the same identifier. Each static object is valid for one interval.

Storage and retrieval of spatio-temporal objects in a DBMS needs more abstractions beyond those discussed in Sect. 12.3.2. TerraLib distinguishes four types of data stored in layers, namely *static* (unchanging data), *events* (singular occurrences such as crime events), *moving objects* (such as cars on highways) and *evolving objects* (such as cities, whose boundaries and attribute values change in time). All spatio-temporal objects that share the same attributes are converted to tuples of a layer (including timestamps) and stored together in a database. Metadata tables store information about different types of layers and identify which attributes of a layer store the timestamps associated to the temporal intervals. For example, consider an urban cadastre where all land parcels in a city for all intervals are stored together in a single layer. Grouping all objects together in this manner simplifies data handling. Inside the database, TerraLib uses optimization techniques for dealing with large data volumes.

As discussed in Sect. 12.3.2, a GIS application needs to transform database tuples (data sources) into objects that can be manipulated and displayed (data targets). In a purely static and non-temporal GIS, it is enough to use *themes* to group objects of the same type resulting from queries. In this case, all objects contained in a theme belong to the same interval.

When a *theme* of spatio-temporal data is retrieved from the database, it contains all spatio-temporal objects for the whole period when data are available. However, not all objects exist in all instances. Consider the case of real estate in an urban cadastre. Extracting data from the "parcels" *layer* will produce a *theme* composed of all parcels that ever existed in the cadastre. An application may need to use only those parcels that currently exist. In this case, a spatio-temporal selection is required inside a *theme*. To do this, TerraLib uses an extra concept referred to as a spatio-temporal object (*STObject*). An *STObject* is an individual entity that preserves its identity, but may change its location and the values of its attributes.

TerraLib provides a query processor to extract *STObjects* from themes (Ferreira et al. 2005, see Fig. 12.3). The query processor has a generic API for programmers, hiding data storage details. Algorithms can handle spatio-temporal data using only *STObjects* returned by the query processor. To perform a query, a programmer defines three different restrictions (spatial, temporal, and attribute). The spatial restrictions use the OGC-specified topological predicates and the temporal restrictions use Allen's interval predicates: *before, meets, overlaps, finished, during, starts*, and *equals* (Allen 1983). Using combinations of these restrictions, the query processor is able to respond to questions such as: *"For each month, which changes occurred in the parcel?"*, *"Which crimes happened on Friday in the south zone of Rio de Janeiro?"* and *"What was the path followed by this wolf in July of 2007?"*

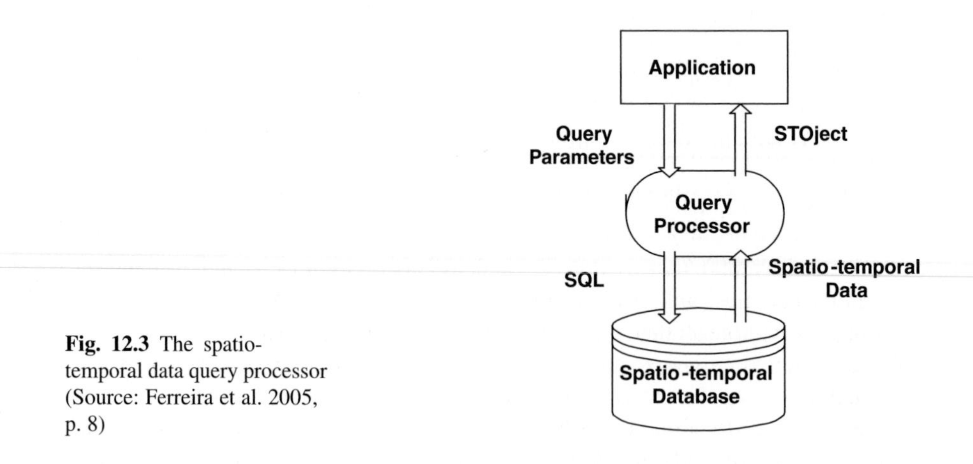

Fig. 12.3 The spatio-temporal data query processor (Source: Ferreira et al. 2005, p. 8)

12.3.6 Spatial Statistics in TerraLib

A GIS produces colour maps of variables such as individual counts quality of life indexes, or company sales in a region. However, to explore the underlying information present in the data, visualisation is not enough. To make effective use of environmental and socioeconomic data, a GIS should provide statistical methods and models. *Spatial statistical* methods measure properties and relations and translate the existing patterns into objective measures. They include geostatistics (Goovaerts 1997), global and local autocorrelation indexes (Anselin 1995), analysis of point patterns (Diggle 2003), regionalization (Openshaw and Alvanides 2001; Martin 2003) and spatial regression (Anselin 1988).

One approach to link spatial statistics to GIS is to use loose coupling mechanisms, where the GIS does data conversion and graphic display, and the spatial models run separately. Examples of this approach are the links between SpaceStat and ArcView (Anselin and Bao 1997) and between R and GRASS (Bivand and Neteler 2000). A more recent trend is to integrate spatial statistics methods directly into a GIS. An example of this is GeoDa (Anselin et al. 2006), where a graphical user interface (GUI) is provided for exploratory spatial data analysis on points and polygons.

TerraLib has a basic spatial statistical package, including local and global autocorrelation indexes, non-parametric kernel estimators, and regionalization methods (Assunção et al. 2006). These functions can be used by GIS applications. One such application is TerraView, described in the next section. Additionally, TerraLib provides a direct link with the R programming language using the *aRT* package (Andrade and Ribeiro 2005). R is an open source programming language for statistical computing and graphics and has become a *de facto* standard for developing statistical software (Ihaka and Gentleman 1996).

R has contributors from all over the world, with continuous improvement that incorporates cutting-edge statistical methods. Integration with R can keep a GIS always updated with recent research on spatial statistics. Packages in R relevant to GIS include *geoR* for geostatistics (Diggle and Ribeiro 2007), *splancs* for analysis of point processes (Rowlingson and Diggle 1993), and *sp* that provides general support for spatial analysis (Pebesma and Bivand 2005).

The *aRT* API performs spatial queries and operations in R. It encapsulates TerraLib functions into R objects, and enables R users to read data from a TerraLib database. This coupling satisfies three requirements:

(a) Statisticians can implement methods of data analysis using R and call TerraLib's facilities for data storage and computational geometry directly for R;
(b) TerraLib programmers can quickly develop interfaces for calling wrapped R code, which consists of functions and a description of their arguments. They do not need to know about R internals;
(c) Users of TerraLib-based applications can perform data analysis in R, without knowing R syntax or even without noticing their analysis is executed by R.

An example of the R-TerraLib coupling is shown in Fig. 12.4. The data in this case are a set of point samples stored in a TerraLib database. These data were interpolated into a grid using the *geoR* package (Diggle and Ribeiro 2007) and the result stored as a TerraLib layer and displayed using the TerraView Version 3.0 GIS application.

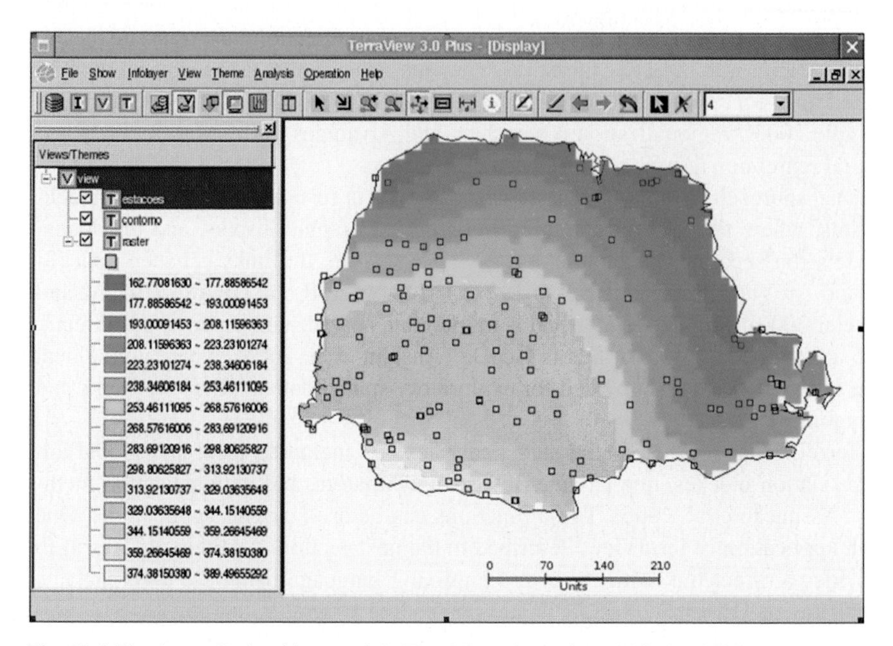

Fig. 12.4 Plotting an R algorithm result in TerraView (Andrade and Ribeiro 2005)

12.3.7 Cell Spaces and Cellular Automata

A cell space is a spatial data type where each cell handles one or more types of attribute. Cell spaces were part of early GIS implementations (Dutton 1978), and later discarded by one-attribute raster data structures, mostly because of efficiency issues. It is time to reconsider this decision and to reintroduce cell spaces as a basic data type in a GIS. Cell-spaces have several advantages over raster-based layers as a means of storing information about continuous spatial entities. Using one-attribute raster data to store results for dynamical models requires storing information in different files. This separation results in increased complexity in data management and user interface design. A cell-space stores all attributes of a cell together, with significant benefits for modelling in contrast to the more cumbersome single value raster approach.

Cell-spaces have been used in the last two decades for simulation of urban and environmental models as part of cellular automata (CA) models (Batty 2000). Most CA models link to a GIS by loose coupling mechanisms, where the GIS performs the data handling and graphic display, and the spatial models run outside the GIS database. This requires extra work for data translation, and may introduce problems of redundancy and consistency. Modeling tools also lack GIS spatial analytical capacities. To address these drawbacks, cell-space models need strong links to the GIS architecture. Using strong coupling, modeling and GIS can be made more robust through their linkage and co-evolution (Parks 1993).

TerraLib supports cell spaces as one of its native data types. It provides functions for storage and retrieval of cell spaces in a DBMS and algorithms for creating cell spaces from vector data. The use of cell spaces enabled development of the TerraME language, which is an add-on to TerraLib that enables simulation in 2D cellular spaces. It supports multi-scale spatial models, where each scale has a different extent and resolution (Carneiro 2006).

Two important innovations in TerraME are its use of anisotropic spaces (Aguiar et al. 2003) and hybrid automata models (Henzinger 1996). Anisotropic spaces arise when modeling natural and human-related phenomena. For example, land settlers in a new area do not occupy all places at the same time. They follow roads and rivers, leading to an anisotropic pattern. However, most spatial statistical and dynamic modeling techniques fail to incorporate spatial anisotropy, leaving out spatial relations that are variable over space. This leads to a serious challenge in producing models that approximate reality, since most real-life spaces are anisotropic.

A *hybrid automaton* is an abstract model for a system whose behaviour has discrete and continuous parts (Henzinger 1996). It extends the idea of finite automata to allow continuous change to take place between transitions. Adopting hybrid automata in spatial dynamic models allows complex models which include critical transitions. Inside each discrete state, the model variables can change. When a critical value occurs, the model moves from the current state to a new one, which is governed by different equations. For example, consider a model for tropical vegetation that has a critical threshold caused by land use change. Under conditions of small land change, the vegetation follows one growth model. When a critical condition is reached, a different growth model must be used. The use of a hybrid automaton allows modeling the tropical vegetation under two different conditions (Carneiro 2006).

Among the typical applications of TerraME are land change and hydrological models. Figure 12.5 shows an application of TerraME for land use change modeling in the Brazilian Amazon. The scenario considers the possible impact of paving a road between the cities of Porto Velho and Manaus. This road crosses areas of currently pristine tropical forest. The model results provide an estimate of how much increase in deforestation could occur in the region from 1997 to 2020 (Aguiar 2006). These models have proven to be useful for supporting public policies that protect the environment and aim at establishing sustainable development practices.

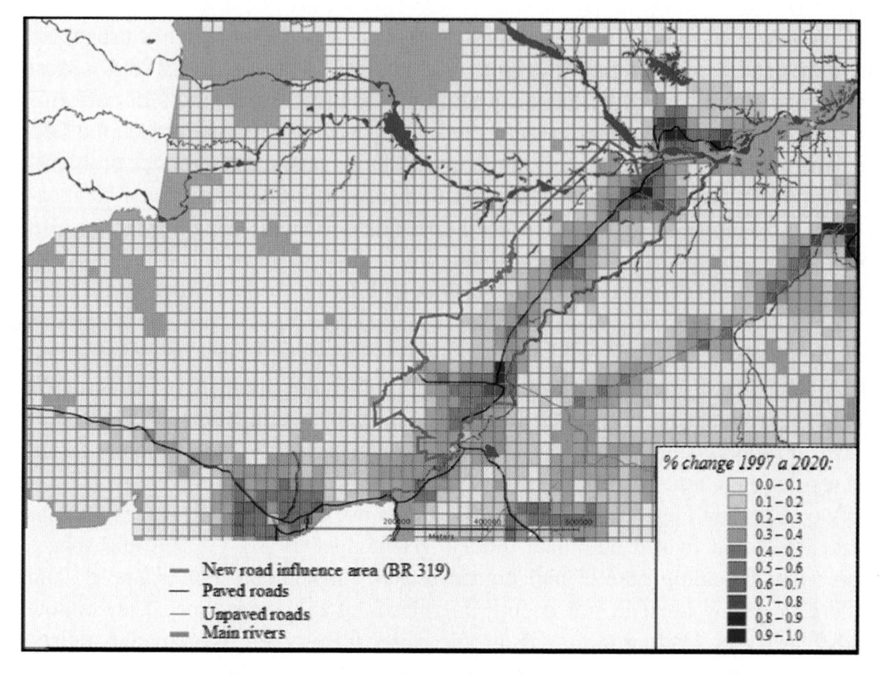

Fig. 12.5 Spatial modeling of projected deforestation of a new road in Amazonia from 1997 to 2020 (Aguiar 2006)

12.4 Development of GIS Applications using TerraLib

This section provides an outline on how to develop a GIS application using TerraLib and describes selected GIS applications that use TerraLib. The general principles of GIS application development are described followed by a description of TerraView, a FOSS4G GIS for spatial data analysis, and TerraAmazon, Brazil's national database for monitoring deforestation.

12.4.1 Building an Application Using TerraLib

TerraLib provides C++ classes that support the higher-level abstractions described in Sect. 12.3, such as *Database*, *Layer*, *View*, and *Theme*. When a programmer builds an application using TerraLib, he/she should use these classes. This section provides a brief guide to the steps involved in GIS application development. TerraLib classes are denoted using `monospaced` font (e.g., `Database` is the TerraLib class for a spatial database). For more detail, see TerraLib's programming tutorial (Vinhas et al. 2007).

Consider first the case of a GIS application using static data, which has four steps: (a) querying the spatial database; (b) converting the query results (tuples) into objects; (c) manipulating these objects to create new objects; and (d) displaying the resulting objects. To perform these steps in TerraLib, a programmer should include code that does the following:

1. Choose the DBMS that will support the application.
2. Create a TerraLib `Database`.
3. Connect to the `Database`.
4. Import data to create a new `Layer` from standard spatial data formats.
5. Create a view to store a user's view of the database using `View`.
6. Create a `Theme` and insert it to the user's `View`.
7. Define the contents of the `Theme`, by pointing to the data source (a `Layer`) and defining attribute and spatial restrictions over that `Layer`.
8. Load the contents of the `Theme`.
9. Define the display parameters of the `Theme` using a `Visual`.
10. Display the `Theme` using a GUI toolkit.

A second situation arises when the GIS application uses spatio-temporal data. In this case, following step 8 above, the developer should include code for the following operations:

9a. Create a `Querier` (query processor).
10a. Define the spatial, attribute and temporal restrictions to apply the query using the `Querier`.
11a. Apply the query and get an `STObjectSet` (set of spatio-temporal objects).
12a. Manipulate and display the `STObjectSet`.

These steps show that Terralib abstractions encapsulate a general view of how a GIS works. The abstractions of *database*, *layer*, *theme*, and *view* are especially important, as they provide a link between what is stored in a database and what is selected and manipulated by a user. Thus, a *database* organizes spatial data in *layers* and a user manipulates *themes* according to his/her *view* of the database. Layers store different types of spatio-temporal data, and thus provide the containers needed by the database.

Although these concepts are not part of the current OGC specifications, they arise from decades of experience. Most commercial and FOSS4G applications use them implicitly or directly. Building a user interface on top of these abstractions is simple. A GUI creates events that match to actions that call TerraLib functions. If the user knows the TerraLib ideas, each of these actions will consist of small pieces of code, based on standard examples.

12.4.2 Examples of Open Source Applications

TerraView is a FOSS4G GIS for spatial data analysis, which provides the basic functions of data conversion, display, exploratory spatial data analysis, spatial

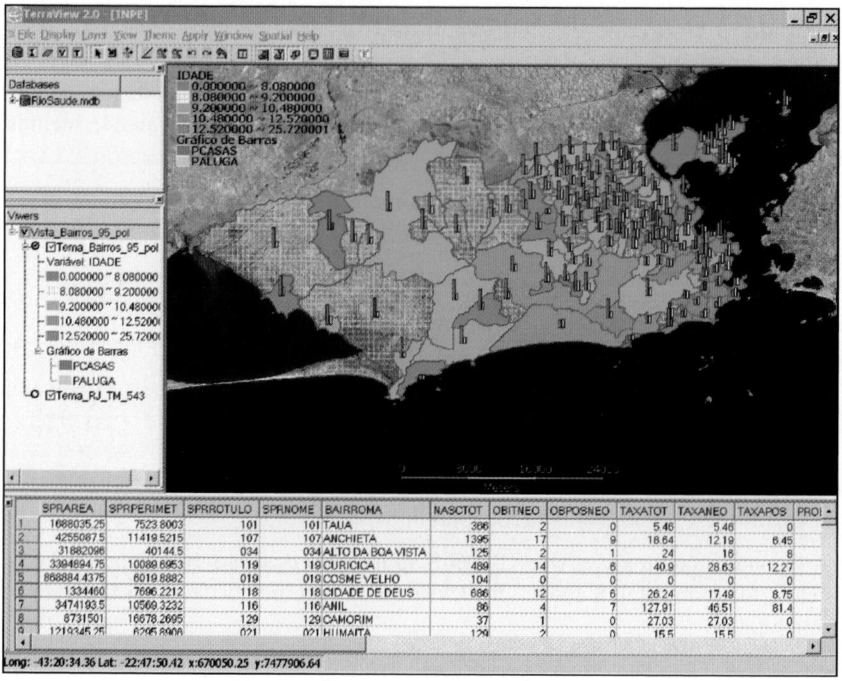

TerraView 2.0 - [INPE]

File Display Layer View Theme Apply Window Spatial Help

Databases

RioSaude.mdb

Viewers

Vista_Bairros_95_pol

Tema_Bairros_95_pol

Variável IDADE

0.000000 ~ 8.080000

8.080000 ~ 9.200000

9.200000 ~ 10.480000

10.480000 ~ 12.520000

12.520000 ~ 25.720000

Gráfico de Barras

PCASAS

PALUGA

Tema_RJ_TM_543

IDADE

0.000000 ~ 8.080000

8.080000 ~ 9.200000

9.200000 ~ 10.480000

10.480000 ~ 12.520000

12.520000 ~ 25.720001

Gráfico de Barras

PCASAS

PALUGA

	SPRAREA	SPRPERIMET	SPRROTULO	SPRNOME	BAIRROMA	NASCTOT	OBITNEO	OBPOSNEO	TAXATOT	TAXANEO	TAXAPOS	PROI
1	1688035.25	7523.8003	101	101	TAUA	366	2	0	5.46	5.46	0	
2	4255087.5	11419.5215	107	107	ANCHIETA	1395	17	9	18.64	12.19	6.45	
3	31882096	40144.5	034	034	ALTO DA BOA VISTA	125	2	1	24	16	8	
4	3394894.75	10089.6953	119	119	CURICICA	489	14	6	40.9	28.63	12.27	
5	868884.4375	6019.8882	019	019	COSME VELHO	104	0	0	0	0	0	
6	1334460	7696.2212	118	118	CIDADE DE DEUS	686	12	6	26.24	17.49	8.75	
7	3474193.5	10569.3232	116	116	ANIL	86	4	7	127.91	46.51	81.4	
8	8731501	16678.2695	129	129	CAMORIM	37	1	0	27.03	27.03	0	
9	1219345.25	6295.8906	021	021	HUMAITA	129	2	0	15.5	15.5	0	

Long: -43:20:34.36 Lat: -22:47:50.42 x:670050.25 y:7477906.64

Fig. 12.6 User interface for the TerraView product

statistical modeling, and spatial and non-spatial queries. The project is a general-purpose GIS for TerraLib databases. Many Brazilian public institutions use TerraView for public policy making, including studies in spatial epidemiology and crime analysis. Figure 12.6 shows TerraView's user interface.

TerraView is licensed using the GNU General Public License (GPL), and is available at http://www.dpi.inpe.br/terraview/. Its user interface uses the QT cross-platform framework (Blanchette and Summerfield 2006). Programmers can extend its functionalities in two ways. By adapting the menus, they can include new functions or change the behaviour of existing ones. Additionally, TerraView supports plug-ins, which are independent applications that can access a TerraLib database. Using TerraLib, plug-ins have access both to the database and to the display controls (i.e., the lists of views, themes, and the canvas).

A second noteworthy application is *TerraAmazon*, Brazil's national database for monitoring deforestation in Amazonia, developed by INPE and its partner, the Foundation for Space Science, Technology and Applications (FUNCATE) (http://www.dpi.inpe.br/terraamazon/). The DBMS is PostgreSQL version 8.2, hosted on a server running the Linux operating system. The application manages all data workflow by gathering about 600 satellite images and pre-processing, segmenting, and classifying these images for further human interpretation in a concurrent multi-user environment (see Fig. 12.7). The database stores about 2 million complex polygons and grows yearly with 60 gigabytes of full resolution satellite images using

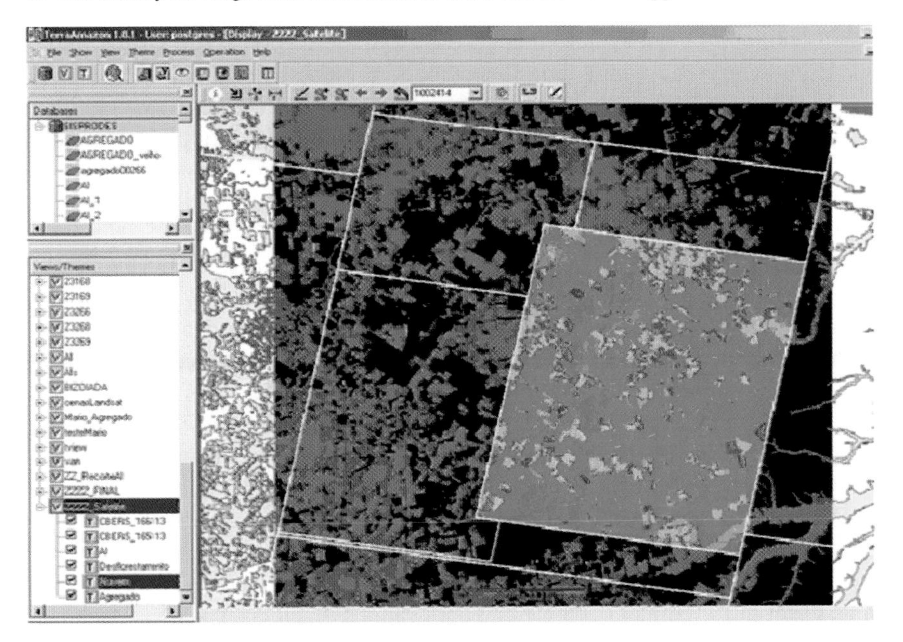

Fig. 12.7 User interface for TerraAmazon

the TerraLib pyramidal resolution schema. A Web site allows seamless display and analysis of full resolution data using TerraLib's PHP extension and TerraLib's OGC WMS server.

12.5 Licensing and Maintenance Policy

One of the important decisions on the TerraLib project was to decide on its license and long-term maintenance policy. The decision considered the nature of the GIS market and the strategy for open source technologies to reach a critical mass of users. The GIS software market is an oligopoly where ESRI, Bentley, and Intergraph have a market share greater than 50% (Daratech 2006). This leads to a "vendor lock-in" effect (Arthur 1994). The "lock-in" effect occurs when a customer is dependent on a vendor for products and services, and cannot move to another vendor without large switching costs. There are many causes for vendor dependence and reluctance to use FOSS4G. First, commercial GIS products use proprietary data formats, making users apprehensive to the costs of data conversion. Second, each GIS adopts a different data model and user interface, which requires training for effective use. Finally, users worry about long-term maintenance of their archives and applications, as the sustainability of FOSS4G projects is often unknown. This results in a conservative policy for most GIS adopters.

Service providers based on OSS face a tough challenge. Convincing a user to change from a commercial to an OS product requires a substantial effort. Users will consider carefully the risks involved in choosing an OS solution, compared to well-publicized commercial products. Software cost is only part of the problem. Users worry about long-term assurance to protect their investments in data capture and in specialized applications. To convince prospective customers, service providers using a FOSS4G approach need to build custom applications fast and reliably, and this is a task that needs investment. Most service providers consider these applications as IP that needs protection. Thus, they are unwilling to invest in OSS that has binding licenses, such as the GNU GPL.

When deciding on the TerraLib license, the developers considered there should be a strong incentive for commercial companies to use the library to reduce the "vendor lock-in" effects of the GIS market in Brazil. Thus, it was decided to release TerraLib according to the GNU Lesser General Public License (LGPL). The LGPL allows private companies to build proprietary applications on top of TerraLib, and market them as proprietary software, while the TerraLib software itself remains publicly licensed. A second consideration involves development and maintenance of the TerraLib kernel. The Brazilian government has guaranteed its long-term support for the core team of developers. INPE provides capacity building for developers, and supports service companies that use the software.

At the time of writing, there are approximately 10 Brazilian service providers using TerraLib for building commercial applications. Each company has a market focus that includes utilities, the oil industry, agriculture, urban cadastre, and the military. There is evidence that this strategy is paying off. Companies offering GIS services based on TerraLib form 10% of the service provider market in Brazil. This impact on the commercial market is an indicator of a decrease in the "vendor lock-in" effect. The library's licensing and maintenance policy are arguably an essential part of this result.

An example of a proprietary application that uses TerraLib is InfoPAE, developed by the Computer Graphics Group (TecGraf) of the Catholic University in Rio de Janeiro (PUC-RIO) in partnership with Petrobras (Petróleo Brasileiro S.A.), a Brazilian oil company. The application was designed for emergency response within the oil industry. InfoPAE works with local emergency action plans (LEAPs) that handle significant events. A LEAP is an organized collection of actions, similar to a workflow, coupled with information stored in geographical as well as conventional databases. LEAP frameworks are useful to design large emergency plans. InfoPAE is being used in more than 100 installations of Petrobras in Brazil.

12.6 Conclusion

The design and implementation of TerraLib serve as an example of the challenges involved in building a FOSS4G library that allows innovative applications and supports large-scale applications. One of main lessons learned is that the current set of OGC specifications is not enough to support these goals. Thus, TerraLib has introduced extra abstractions to reduce the cognitive distance between GIS developers and the outcomes of their work. These extra abstractions come at a price. When a FOSS4G developer adheres strictly to the OGC specifications, his/her code runs in all OGC-compliant libraries. However, adopting TerraLib reduces GIS application development effort, while increasing the cost of using abstractions not supported by other products. OGC-compliant applications will be able to access the part of a TerraLib database that contains OGC's simple features. However, the extra relations used by TerraLib to handle abstractions such as *theme* and *view*, and data structures such as *cell spaces* are invisible to these applications.

Several choices had to be made when introducing solutions for spatio-temporal queries, cell spaces, and raster data handling. Only further experience will show if these were the correct decisions. Another difficult issue concerns the software architecture and design for extensibility. The extensive use of design patterns in TerraLib suits experienced programmers who are comfortable with ideas such as "factory" and "strategy". Novice developers need at least six months training in C++ before they can become skilful in these concepts.

To conclude, developing TerraLib has shown how difficult it is to design a GIS library that combines simplicity and expressiveness. The developers opted for expressiveness at the expense of simplicity in this case. This choice may limit the rate of adoption of TerraLib by the FOSS4G community. Nevertheless, it is the developers' hope the library's assets may be attractive to other developers that want to build large-scale GIS applications.

Acknowledgments *TerraLib's* core team, apart from the authors, includes Laercio Namikawa and Emiliano Castejon at INPE. *TerraView* was designed and implemented by Juan Pinto de Garrido and Lauro Hara. Additional contributors to TerraLib include Tiago Carneiro, designer of *TerraME*, Pedro Andrade, who developed *aRT*, Ana Paula Aguiar, who wrote code for cell spaces, and Felipe Castro da Silva and Thales Korting, who developed image processing functions. The technical support of Julio D'Alge for cartographical projections and Leila Fonseca for image processing has been important. We also have important contributions from Paula Frederick, Marcelo Metello, Natacha Barroso, and Leone Pereira Masieiro at PUC-Rio, and Rui Mauricio Gregório and Vanildes Ribeiro at FUNCATE. The TerraLib project is partially financed by CNPq grant no. 552040/02-9. Gilberto Câmara's research is also financed by CNPq grant no. 300557/96-5. The project has also received financial support from FAPESP (Fundação de Amparo à Pesquisa no Estado de São Paulo).

TerraLib code relies on a number of OSS packages. INPE and TerraLib development team thanks the OS community for their efforts. These libraries are: (a) the *zlib* library to compress data when storing raster data in a TerraLib database; (b) the independent JPEG Group's library for JPEG image compression; (c) *libgeotiff* to decode/encode raster data in TIFF/GEOTIFF format; (d) *shapelib* to decode/encode vector data in shapefile format.

As of late-2007, TerraLib consists of 280,000 lines of C++ code and 170,000 lines of third-party open source utilities. The development started in 2001, with an effort of 60 man-years spent so far in TerraLib and TerraView. The library and associated applications may be obtained from the website http://www.terralib.org.

References

Aguiar A, Câmara G, Cartaxo R (2003) Modeling spatial relations by generalized proximity matrices. V Brazilian Symposium in Geoinformatics – GeoInfo 2003, Campos do Jordão, SP, Brazil

Aguiar APD (2006) Modeling Land Use Change in the Brazilian Amazon: Exploring Intra-Regional Heterogeneity. PhD Thesis, Remote Sensing Program. Sao Jose dos Campos, INPE

Aksoy S, Koperski K, Tusk C, Marchisio G (2004) Interactive training of advanced classifiers for mining remote sensing image archives. ACM International Conference on Knowledge Discovery and Data Mining, Seattle, WA, ACM

Alexandrescu A (2001) Modern C++ design: Generic programming and design patterns applied. Addison-Wesley, Reading

Allen JF (1983) Maintaining knowledge about temporal intervals. Commun ACM 26:832–843

Almeida CM, Batty M, Monteiro AMV, Câmara G, Soares-Filho BS, Cerqueira GC, Pennachin CL (2003) Stochastic cellular automata modeling of urban land use dynamics: empirical development and estimation. Comput Environ Urban Syst 27:481–509

Anselin L (1988) Spatial econometrics: methods and models. Kluwer, Dordrecht

Anselin L (1989) What's special about spatial data: Alternative perspectives on spatial data analysis. Santa Barbara, CA, NCGIA Report 89-4

Anselin L (1995) Local indicators of spatial association – LISA. Geogr Anal 27:91–115

Anselin L, Bao S (1997) Exploratory spatial data analysis linking Spacestat and ArcView. In M Fischer and A Getis (eds.) Recent Developments in Spatial Analysis, Springer Verlag, Berlin.

Anselin L (1999) Interactive techniques and exploratory spatial data analysis. In: Longley P, Goodchild M, Maguire D, Rhind D (eds) Geographical Information Systems: principles, techniques, management and applications. Geoinformation International, Cambridge

Anselin L, Syabri I, Kho Y (2006) GeoDa: An introduction to spatial data analysis. Geogr Anal 38:5–22

Andrade PR, Ribeiro PJ (2005) A process and environment for embedding the R Software into TerraLib. VII Brazilian Symposium on Geoinformatics (GeoInfo 2005), Campos do Jordao, Brazil, INPE/SBC

Arthur B (1994) Increasing returns and path dependence in the economy. The University of Michigan Press, Ann Arbor, MI

Assunção R, Neves M, Camara G, Freitas CDC (2006) Efficient regionalisation techniques for socio-economic geographical units using minimum spanning trees. Int J Geogr Inf Sci 20:797–812

Austern M (1998) Generic programming and the STL: Using and extending the C++ standard template library. Addison-Wesley, Reading, MA

Batory D, Singhal V, Sirkin M, Thomas J (1993) Scalable software libraries. SIGSOFT Softw. Eng. Notes 18:191–199

Batty M (2000) GeoComputation using cellular automata. In: Openshaw S, Abrahart RJ (eds) GeoComputation, Taylor & Francis, London, 95–126

Blanchette J, Summerfield M (2006) C++ GUI programming with Qt 4. Prentice Hall, Indianapolis, Indiana

Blaschke T, Hay G (2001) Object-oriented image analysis and scale-space: theory and methods for modeling and evaluating multiscale landscape structure. Int Arch Photogramm Remote Sens 34:22–29

Bivand R, Neteler M (2000) Open source geocomputation: using the R data analysis language integrated with GRASS GIS and PostgreSQL data base systems. 5th International Conference on GeoComputation, Greenwich, UK

Burrough P (1998) Dynamic modelling and geocomputation. In: Longley P, Brooks S, McDonnell R, Macmillan B (eds) Geocomputation: A Primer. John Wiley, New York

Câmara G, Souza R, Freitas U, Garrido J (1996) SPRING: Integrating remote sensing and GIS with object-oriented data modelling. Comput Graph 15:13–22

Câmara G, Souza RCM, Pedrosa BM, Vinhas L, Monteiro AMV, Paiva JAC, Carvalho MT, Raoult B (2001) Design patterns in GIS development: the TerraLib experience. III Simpósio Brasileiro de GeoInformatica, Rio de Janeiro, RJ

Carneiro T (2006) Nested-CA: a foundation for multiscale modeling of land use and land change. Computer Science Department. Sao Jose dos Campos, INPE. Doctorate Thesis in Computer Science

Chang SK, Yan CW, Dimitroff D, Arndt T (1988) An intelligent image database system. IEEE Trans Software Eng 14:681–688
Couclelis H (1997) From cellular automata to urban models: New principles for model development and implementation. Environ Plann B 24:165–174
Daratech (2006) GIS markets and opportunities 2006 survey. Cambridge, MA, Daratech Inc
DeWitt D, Kabra N, Luo J, Patel J, Yu J-B (1994) Client-server paradise. VLDB Conference, Santiago, Chile
Diggle P (2003) Statistical analysis of spatial point patterns. 2nd edn Edward Arnold, London
Diggle P, Ribeiro PJ (2007) Model-based geostatistics. Springer, Heidelberg
Dutton G (ed) (1978) First international advanced study symposium on topological data structures for geographic information systems. Addison-Wesley, Reading, MA
Elliott J, Eckstein R, Loy M, Wood D, Cole B (2002) Java swing. O'Reilly Press, Sebastopol, CA
Erwig M, Schneider M (2002) Spatio-temporal predicates. IEEE Trans Knowl Data Eng 14:881–901
Feitosa F, Camara G, Monteiro AM, Koschitzki T, Silva MS (2007) Global and local spatial indices of urban segregation. Int J Geogr Inf Sci 21:299–323
Ferreira KR, Queiroz G, Paiva JA, Souza RC, Câmara G (2002) A software architecture for building spatial databases with object-relational DBMS. XVII Brazilian Symposium on Databases, Gramado, RS
Ferreira KR, Vinhas L, Queiroz GR, Câmara G, Souza RCM (2005) The architecture of a flexible querier for spatio-temporal databases. VII Brazilian Symposium in Geoinformatics, Campos do Jordao, Brazil
Fonseca F, Egenhofer M, Agouris P, Camara G (2002) Using ontologies for integrated geographic information systems. Trans GIS 6:231–257
Fotheringham AS, Brunsdon C, Charlton M (2002) Geographically weighted regression: The analysis of spatially varying relationships. Wiley, Chichester
Fowler GS, Korn DG, Vo K-P (1995) Principles for writing reusable libraries. Proceedings of the 1995 Symposium on Software reusability. Seattle, Washington, United States, ACM Press
Galton A (2004) Fields and objects in space, time, and space-time. Spat Cogn Comput 4:39–68
Gamma E, Helm R, Johnson R, Vlissides J (1995) Design patterns: Elements of reusable object-oriented software. Addison-Wesley, Reading, MA
Goodchild ME (2003) Geographic information science and systems for environmental management. Ann Rev Environ Resour 28:493–519
Goovaerts P (1997) Geostatistics for natural resources evaluation. Oxford Univ Press, New York
Grenon P, Smith B (2003) SNAP and SPAN: Towards dynamic spatial ontology. Spat Cogn Comput 4:69–104
Güting RH, Schneider M (2005) Moving objects databases. Morgan Kaufmann, New York
Henzinger TA (1996) The theory of hybrid automata. Proceedings of the 11th Symposium on Logic in Computer Science (LICS'96), IEEE
Hornsby K, Egenhofer M (2000) Identity-based change: A foundation for spatio-temporal knowledge representation. Int J Geogr Inf Sci 14:207–224
Ihaka R, Gentleman R (1996) R: A language for data analysis and graphics. J Comput Graph Stat 5:299–314
Krasner GE, Pope ST (1988) A cookbook for using the model-view controller user interface paradigm in Smalltalk-80. J Object-Oriented Program 1:26–49
Krueger CW (1992) Software reuse. ACM Comput Surv 24:131–183
Martin D (2003) Extending the automated zoning procedure to reconcile incompatible zoning systems. Int J Geogr Inf Sci 17:181–196
Medak D (2001) Lifestyles. In: Frank AU, Raper J, Cheylan J-P (eds) Life and Motion of Socio-Economic Units. ESF Series. Taylor & Francis, London
Meinel G, Neubert M (2004) A comparison of segmentation programs for high resolution remote sensing data. Int Arch Photogramm Remote Sens XXXV:1097–1105
Meyer B (1990) Lessons from the design of the Eiffel libraries. Commun ACM 33:68–88

Mockus A, Fielding R, Herbsleb J (2002) Two case studies of open source software development: Apache and Mozilla. ACM Transactions on Software Engineering and Methodology 11

Navulur K (2006) Multispectral image analysis using the object-oriented paradigm. CRC Press, Boca Raton, CA

Openshaw S, Alvanides S (2001) Designing zoning systems for representation of socio-economic data. In: Frank A, Raper J, Cheylan J (eds) Time and Motion of Socio-Economic Units, Taylor and Francis, London

Parks BO (1993) The need for integration. In: Goodchild MJ, Parks BO, Steyaert LT (eds) Environmental modelling with GIS. OUP, Oxford, 31–34

Pebesma E, Bivand R (2005) Classes and methods for spatial data in R. R News 5:9–13

Pedrosa B, Câmara G, Fonseca F, Souza RCM (2002) TerraML – A cell-based modeling language for an open-source GIS library. II International Conference on Geographical Information Science (GIScience 2002), Boulder, CO, 2002

Queiroz GR (2003) Algoritmos Geométricos para Bancos de Dados Geográficos: Da Teoria à Prática na TerraLib (Geometric Algorithms for Spatial Databases: From Theory to Practice in TerraLib). Computer Science. São José dos Campos, INPE. **MsC**

Rodrigues VL, Andrade MVA, Queiroz GR, Magalhães M (2006) An efficient map overlay algorithm for TerraLib. VIII Brazilian Symposium on GeoInformatics, GeoInfo2006, Campos do Jordão, SP, Brazil, INPE

Rodrigues VL, Cavalier AP, Andrade MVA, Queiroz GR (2005) Exact algorithms for map manipulation in TerraLib. VII Brazilian Symposium on GeoInformatics, GeoInfo2005, Campos do Jordão, SP, Brazil, INPE

Rowlingson B, Diggle P (1993) Splancs: spatial point pattern analysis code in S-Plus. Comput Geosci 19:627–655

Silva MPS, Camara G, Souza RCM, Valeriano D, Escada MIS (2005) Mining patterns of change in remote sensing image databases. The Fifth IEEE International Conference on Data Mining, New Orleans, Louisiana, USA

Sistla AP, Wolfson O, Chamberlain S, Dao S (1997) Modeling and querying moving objects. Proceedings of the Thirteenth International Conference on Data Engineering 422–432

Veldkamp A, Fresco L (1996) CLUE: A conceptual model to study the conversion of land use and its effects. Ecol Model 85:253–270

Vinhas L, Ferreira KR, Ribeiro G (2007) TerraLib programming tutorial. São José dos Campos, Brasil, INPE (avaliable on http://www.terralib.org)

Vinhas L, Queiroz GR, Ferreira K, Câmara G, Paiva JA (2002) Generic programming applied to GIS algorithms. IV Brazilian Symposium on Geoinformatics, Caxambu, Brazil

Vinhas L, Souza RCM, Câmara G (2003) Image data handling in spatial databases. V Brazilian Symposium on Geoinformatics, Campos do Jordão, Brazil

Warmerdam F (2007) Shapefile C Library V1.2, http://shapelib.maptools.org/Last accessed July 28th, 2008

Weber S (2004) The success of open source. Harvard University Press, Cambridge, 75

Worboys M (2005) Event-oriented approaches to geographic phenomena. Int J of Geogr Inf Syst 19:1–28

Index

Author Biography

G. Brent Hall completed his PhD in Geography at McMaster University in Hamilton, Ontario, Canada in 1980. He was a Professor of Planning at the University of Waterloo, Ontario, Canada until 2007 when he was appointed as Dean of the National School of Surveying at the University of Otago in Dunedin, New Zealand. His teaching has been devoted to geographic information systems and his recent research focuses on participatory Web-based GIS for decision support using Open Source geospatial components. Dr. Hall has published one textbook, numerous book chapters and over fifty refereed journal papers on various aspects of spatial analysis and GIS.

Michael Leahy was born and raised in London, Ontario, Canada. He graduated from the Geography program at the University of Waterloo with Honours Bachelor of Environmental Studies in 2002, and Masters of Environmental Studies in 2005. He is a Doctoral candidate in the Geography and Environmental Studies program at Wilfrid Laurier University, and he is currently focused on applying FOSS4G technologies for the development of the MapChat tool, and is engaged in related case study work in a rural community in New Zealand for his doctoral thesis.

Printing: Krips bv, Meppel, The Netherlands
Binding: Stürtz, Würzburg, Germany